Steam Turbin

SECOND EDI

G. J. Roy, CEng, FIMarE, MRINA

Foreword by
W. D. Ewart, CEng, FIMarE, MRINA

STANFORD MARITIME LONDON

Stanford Maritime Limited
Member Company of the George Philip Group
12–14 Long Acre, London WC2E 9LP

First published 1975
Reprinted 1978
Second edition 1984
Reprinted 1987

Printed in Great Britain by
J. W. Arrowsmith Limited Bristol

British Library Cataloguing in Publication Data

Roy, G.J.
Steam-turbines and gearing—2nd ed.
1. Steam-turbines, Marine
I. Title
623.8'722 VM740

ISBN 0 540 07358 X

Foreword

Written by experienced lecturers at one of Britain's leading marine engineering colleges, each book of this series is concerned with a subject in the syllabus for the examination for the Second Class Certificate of Competency. It is intended that the books should supplement the standard text books by providing engineers with numerous worked examples as well as easily understood descriptions of equipment and methods of operation. Extensive use is made of the question and answer technique and specially selected illustrations enable the reader to understand and remember important machinery details.

While the books form an important basis for pre-examination study they may also be used for revision purposes by engineers studying for the First Class Certificate of Competency.

Long experience in the operation of correspondence courses has ensured that the authors treat their subjects in a concise and simple manner suitable for individual study—an important feature for engineers studying at sea.

W. D. Ewart

Preface to the Second Edition

The aim of this book is to provide basic information on steam turbines, gearing and associated equipment as used at sea which is related to the knowledge required by a student studying for the Class Two Certificate of Competency as a Marine Engineer Officer issued by the examining body responsible for these certificates in the United Kingdom.

It is the intention of this book to give a guide to the subject, and to provide a foundation upon which the prospective candidate, with his own experience, can provide suitable examination answers. The Second Edition has been updated in line with current common practice.

The diagrams in this book are not drawn to scale, and indicate the type of information that should be given, although in the examination there will not be time to attain the same detail.

In a few cases pictorial representations have been used to show components more clearly, but unless the candidate has some reasonable ability at this type of sketching, it should as a rule be avoided in the examination.

The sketches required in the examination need not be drawn to scale, but should be in proportion unless in diagrammatic form, this being allowed provided the principles concerned are clearly indicated.

Drawing instruments may be used, but in general their use will take up valuable time, and as much of the sketch as possible should be freehand. Colours may be used provided they do not confuse the completed sketch.

I am indebted to the following manufacturers for permission to reproduce diagrams of their equipment:

GEC Turbine Generators Limited of Manchester—Fig. 33;
Peter Brotherhood Limited of Peterborough—Figs. 42 and 43;
Stal Laval Limited–Figs. 8(a,b,c), 15, 27(b), 30, 34, 37, 39, 51, 52, 54.

G. J. Roy

Contents

SI UNITS

Mass	= kilogramme	(kg)
Force	= newton	(N)
Length	= metre	(m)
Pressure	= newton/sq metre	(N/m^2)
Temperature	= degrees celcius	(°C)

CONVERSIONS

1 inch = 25·4mm = 0·025m
1 foot = 0·3048m
1 square foot = 0·093m^2
1 cubic foot = 0·028m^3
1 pound mass (lb) = 0·453kg
1 UK ton (mass) = 1016kg
1 short ton (mass) = 907kg
1 tonne mass = 1000kg
1 pound force (lbf) = 4·45N
1 ton force (tonf) = 9·96kN
1kg = 9·81N

0·001in = 0·025mm
(°F − 32) × = °C
1lbf/in^2 = 6895N/m^2 = 6·895kN/m^2
1kg/cm^2 = 1kp/cm^2 = 102kN/m^2
1 atmos = 14·7lbf/in^2 = 101·35kN/m^2
1 bar = 14·5lbf/in^2 = 100kN/m^2
Note: For approximate conversion of pressure units
100kN/m^2 = 1 bar = 1kg/cm^2 = 1 atmos
1 tonf/in^2 = 15440kN/m^2 = 15·44MN/m^2
1HP = 0·746kW

CHAPTER 1

Turbine Types

Behaviour of steam flow in impulse and Parsons impulse reaction turbine blades. Differences in cross section.

Fig. 1(a) shows a cross-section through a pair of adjacent impulse turbine blades, whilst Fig. 1(b) shows a cross-section through a pair of adjacent Parsons reaction turbine blades. In both cases use is made of the impulse principle by reducing the kinetic energy of a stream of steam passing across the blades. A change

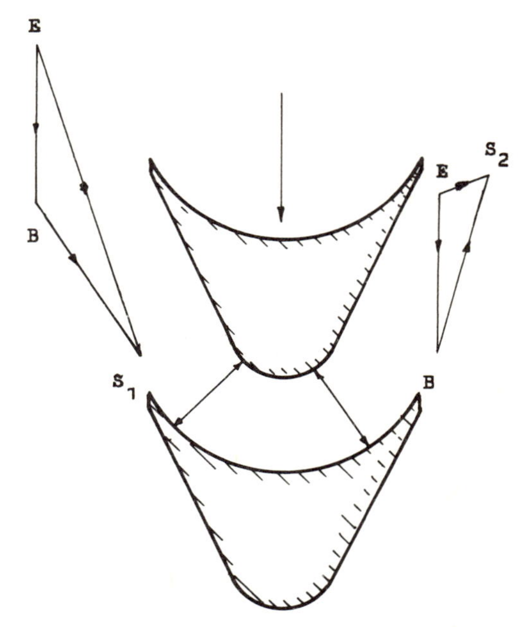

Fig. 1(a) Impulse turbine blades

of kinetic energy is accompanied by a change of velocity, but velocity is a vector quantity indicating speed in a given direction. Change either speed or direction, or both, and the velocity has been changed, with the resulting change in kinetic energy producing a force acting upon the turbine blade to turn the rotor. Again in both cases, the stream of steam is accelerated in a ring of nozzles fixed to the turbine casing immediately prior to passing over the blades. With the impulse turbine the moving blades are designed so that the stream of steam passes through a uniform path between adjacent blades without any pressure drop taking place, the stream of steam only having its direction changed. This change of direction is thus a change of velocity which produces a force acting upon the blade to turn the rotor. In theory the speed of the stream of steam leaving the blade, measured relative to the blade, is the same as that at entry, but in practice, due to friction and eddy losses, the speed at exit tends to drop and there is some re-heating of the steam causing a loss of efficiency.

In the Parsons reaction turbine, the stream of steam enters the blade passage and a force is produced due to a change in direction, as in the impulse turbine. As Fig. 1(b) shows, the design of the blades for this turbine is such that adjacent blades form

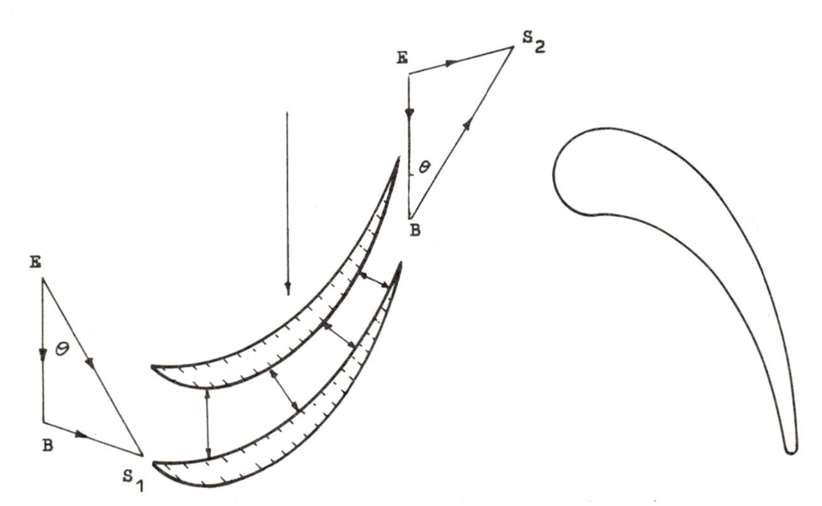

Fig. 1(b) Reaction turbine blades

Fig. 1(c) Pear-drop shaped blade

a passage of reducing dimensions, so that if the same mass of steam is to pass through each point along its length per second, the speed of the stream of steam must increase. The energy to do this is obtained in a conversion of pressure energy into kinetic energy, hence a pressure drop. This increase of speed across the blade produces a change of velocity, thus inducing a force on the blade. Therefore in this turbine, the force on the blade is due to changes in both the direction and speed of the stream of steam, the latter giving the reaction effect which accounts for about 50 per cent of the force on the blade. This pressure drop across the moving blade is the main difference between a Parsons reaction turbine and an impulse turbine, the former being more correctly called an impulse reaction turbine.

In practice impulse turbine blades are designed with a small degree of reaction to overcome the effect of friction re-heat, therefore improving efficiency.

Forces on impulse turbine blades tend to be higher than those on the blades of impulse-reaction turbines due to the smaller pressure drops involved in the latter and therefore the smaller the changes in kinetic energy, hence the difference in blade

cross-section areas. Also impulse turbines tend to run at higher speeds producing greater centrifugal stresses on the blades.
In Fig. 1(a) and Fig. 1(b):

E − B = Velocity of moving blade measured relative to the casing

E − S_1 = Velocity of steam flow onto blade measured relative to the casing

E − S_2 = Velocity of steam flow leaving blade measured relative to the casing

B − S_1 = Velocity of steam flow onto blade measured relative to the moving blade

B − S_2 = Velocity of steam flow leaving blade measured relative to the moving blade

Some modern impulse turbine blades are pear or teardrop shaped (Fig. 1(c)) to improve steam entry conditions, particularly when running at non-ideal operating speeds. Shock and turbulence in the flow of steam are reduced to some extent. Blades towards the exhaust end of the LP turbine may be twisted along their length and also tapered. The former gives a variable blade entry angle to accommodate the difference between the steady linear speed of the steam as it leaves the nozzle and the increasing linear blade speed between root and tip, and thus reduces shock and turbulence. Tapering the blade along its length helps to reduce centrifugal stresses in long blades.

Behaviour of steam in convergent and convergent-divergent nozzles

In order to provide the rotational force to drive a steam turbine a stream of steam gives up kinetic energy to the blades attached to the turbine rotor. Steam leaving the boiler has very little kinetic energy, but it does have a high heat energy content, and for any given pressure and condition, i.e. wet, saturated or superheated, it has a particular value of enthalpy, which is a measure of this. If this pressure and condition is changed by, for example, dropping the pressure, then some of this heat energy will be released. If in this expansion of the steam, this energy is not used to drive any mechanical device, then, by the law of conservation of energy, it must appear in another form. By creating a pressure drop across a specially shaped passage as in a nozzle, this interchange of energy can be controlled and it manifests itself mainly as an increase in kinetic energy. It is assumed that the same mass of steam must flow past any point along the nozzle per unit time. At the inlet to the nozzle, the specific volume of the steam is comparatively low and the rate of increase in volume is also small, but the velocity increases at a greater rate. The cross-section area for the steady flow condition is proportional to the ratio of the specific volume of the steam to its velocity and therefore the area required for flow contracts. As the expansion proceeds, however, the rate of increase of the specific volume suddenly rises, at one particular point, at a greater rate than that of the velocity, and the cross-section area begins to increase. At the point immediately prior to that sudden change in the rate of increase of the specific volume, the minimum cross-section area of the nozzle, known as the throat, occurs. If, for the remainder of the length of the nozzle, the area is then kept constant, the design is known as a convergent nozzle and the steam leaves the nozzle without any discontinuity of flow. The quantity of steam discharged will depend upon the drop between inlet and exhaust pressure, but the maximum quantity the nozzle is capable of discharging occurs when the discharge or exhaust pressure is approximately 0·55 × inlet pressure if the steam is superheated, and approximately 0·58 × inlet pressure if the steam is wet or saturated. The pressure at

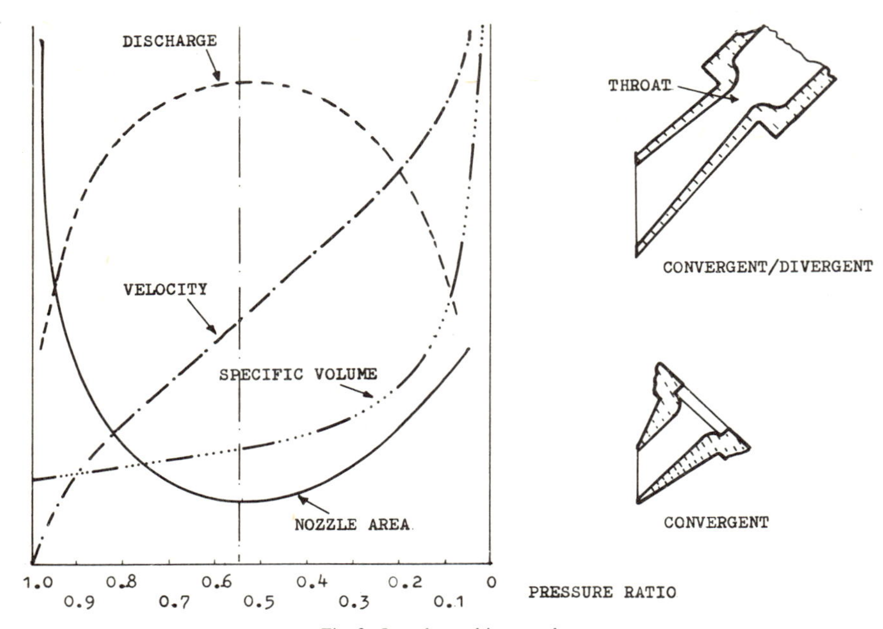

Fig. 2 Impulse turbine nozzles

this particular condition is known as the critical pressure and however low the pressure in the exhaust space is below this critical value, no greater quantity of steam will flow, and the pressure at the throat will never be less than the figures quoted for the critical condition.

If the nozzle has to discharge into a space with a pressure below that for the critical condition, then the instant the steam leaves the nozzle, the expansion becomes uncontrolled, and the volume increases instantaneously with a rapid dissipation of energy, scattering the steam and causing turbulence in the steam flow. To control this expansion under these conditions a divergent section is fitted and should be designed so that the exit area allows just enough controlled expansion to make the discharge pressure equal to the back pressure. With such a convergent-divergent nozzle, the steam leaves the nozzle without any discontinuity of flow, but the maximum quantity flowing is still dependent upon the throat cross-section area. The divergent section is given an angle of divergence of about 8° to 10° either side of the centre line through the nozzle to prevent turbulence, but the converging part is usually made short as the rapid contraction tends to damp turbulence and helps stream line or laminar flow.

In theory the expansion through the nozzle should be adiabatic, as this gives the greatest possible heat drop, but manufacturing defects, carry-over deposits, corrosion and erosion tend to reduce velocity and some of the kinetic energy converts back into heat energy, re-heating the steam. For this reason the greatest care should be taken in cleaning air ejector nozzles, etc., upon which no abrasive material should be used. Fig. 2 shows a convergent nozzle and a convergent-divergent nozzle, together with a graph showing the behaviour of the steam velocity, specific volume, and quantity discharged.

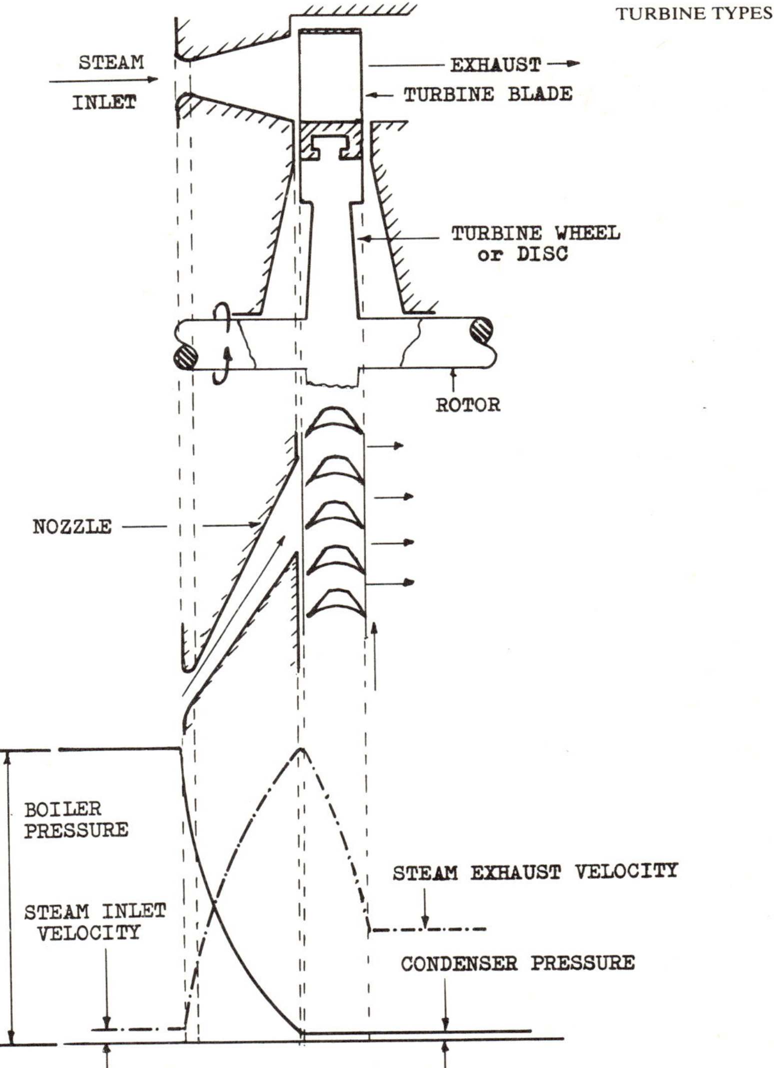

Fig. 3 De Laval impulse turbine also called a single-velocity compounded wheel. See Fig. 9.

Pressure and velocity variations in De Laval turbine. Use in marine applications.

Fig. 3 shows the basic design of such a turbine with an approximate indication of the variation in absolute steam velocity and absolute steam pressure as the steam passes through the turbine. The design consists of a single ring of nozzles directing steam onto a single row of blades attached to a wheel or disc on the rotor. Steam at inlet pressure is admitted to the nozzles, through which it expands down to the pressure at the exit from the blades. This may be condenser pressure or a suitable pressure to match the requirements of the application of the turbine. This exhaust pressure is the same as the exit pressure from the nozzles, since there is no pressure drop across the blades. As the steam expands in the nozzles an interchange of energy takes place: the heat energy released as the pressure drops, converting into kinetic energy, and manifesting itself as an increase in the speed of the steam. On passing across the turbine blades this high speed flow of steam has its direction changed, producing a force on the blades. For the most economical operation the blade speed should be about 0·5 of the steam speed measured relative to the casing. If, for example, the pressure drop in the nozzles released some 500 kilojoules of heat energy, the result of the interchange to kinetic energy would produce a steam jet leaving the nozzle at about 1480 metres per second, requiring a blade speed of 740 metres per second. An average rotor of one metre diameter would have to rotate at well above 15000rpm to give this blade speed, producing excessively high reduction gear ratios to match the low rpm required by the propeller for optimum efficiency. The blade centrifugal stresses and journal speeds would make equally impracticable demands on blade dimensions and materials as well as on bearing materials and lubrication.

The design may be used at the steam inlet end of some HP turbines to give a large initial drop in pressure and temperature, thus shortening the turbine and easing alloying and scantling requirements for the materials for the remaining stages. Only the nozzle box is subjected to steam at superheater outlet temperature, limiting the need for special alloys to this item. Gland and blade tip leakage is also reduced, as is gland length. It will be seen from the velocity diagram that the exhaust velocity of the steam represents a considerable proportion of the maximum absolute velocity of the steam leaving the nozzle. This represents a considerable loss of kinetic energy to the condenser if the turbine is used in this way, but if the steam passes on for further use, then much of this loss can be recovered.

Pressure compounded turbine

Fig. 4 shows the initial stages of a pressure compounded turbine, together with diagrams representing the behaviour of the steam pressure and velocity between inlet and exhaust. The design basically consists of a number of De Laval turbines placed in series on the same shaft, with the exhaust from each turbine, or stage, entering the next nozzle ring along the shaft. The overall pressure drop and hence heat drop between steam inlet and exhaust is thus divided into a number of smaller pressure drops, i.e. compounding the pressure drops. One stage consists of a division plate or diaphragm containing nozzles through which the steam expands, around its periphery, and a disc or wheel carrying blades over which the flow of steam from the nozzles passes. These blades change the direction of the flow of steam, producing a change in velocity and therefore a rotational force on the wheel. The wheels or discs may be an integral part of the rotor, or fitted with a key and/or a shrunk-force fit on a central spindle. As with the De Laval turbine the pressure drop takes place in the nozzle alone, with the space between adjacent diaphragms, in which the blades move, being at constant pressure. In practice a small pressure drop is provided across the blades to overcome the effect of friction between the steam and blade surfaces.

By compounding or splitting the overall pressure drop into a number of smaller expansions, the overall heat energy available for interchange into kinetic energy is also split so that the steam flow speeds produced at the exit from each ring of nozzles can be made acceptable for marine practice. At each stage the blade speed should be

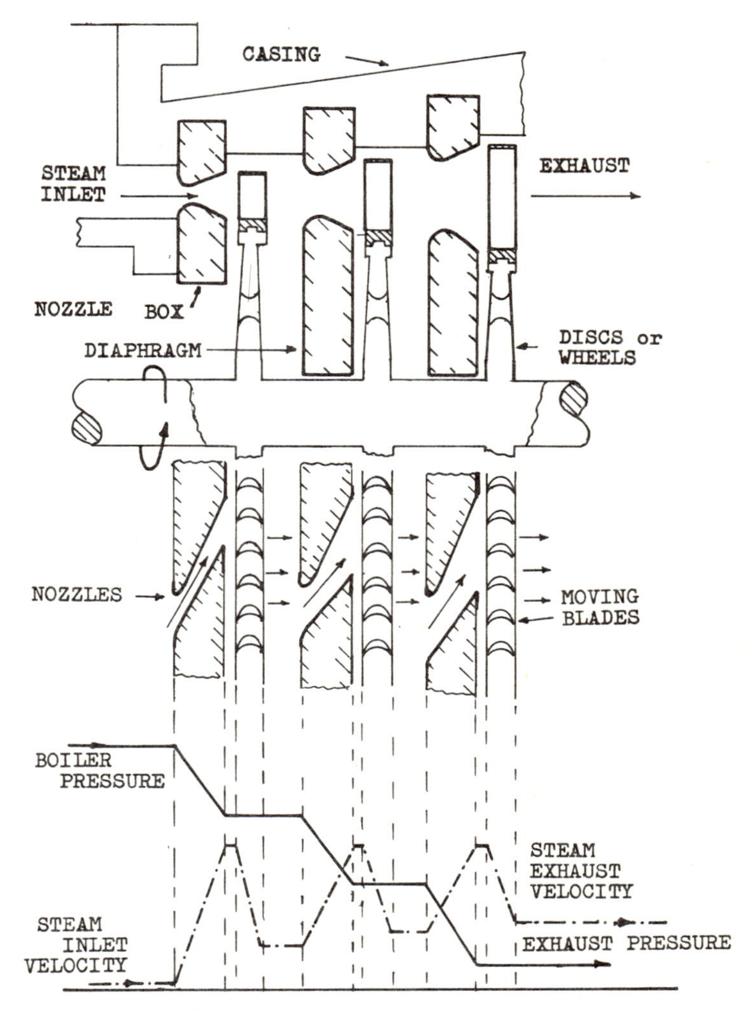

Fig. 4 Pressure compounded impulse turbine

about 0·5 × steam flow speed for maximum efficiency, and therefore by careful design with regard to the number of stages and steam flow speeds, blade mean diameters, speeds, lengths and stresses, rotor rpm combined with suitable gear ratios can be arranged to give the required low propeller rpm, whilst maintaining an acceptable turbine efficiency.

Although it would appear that, by increasing the number of stages, rotor speeds could be reduced to any desired value, this would produce an excessively long and heavy rotor subject to distortion. Stage losses due to steam and disc friction, windage and leakage are also multiplied, although leakage through the annular clearance between the rotor and diaphragm is reduced by labyrinth packing. Stage mean diameters and blade and nozzle height are usually increased as the steam expands to provide an enlarged flow area for the passage of steam at the lower pressure stages but centrifugal stresses have to be watched and blade angles may have to be altered to accommodate the steam flow. This may reduce efficiency, but the stage efficiency is high when compared with other compounding arrangements which have the advantage of shortening the rotor, but at the expense of much lower efficiency. The design can produce a comparatively short, robust turbine, suitable for speeds at the moment in the region of 6000rpm, with an acceptable efficiency suitable for both main propulsion and auxiliary purposes. The design also may be known as a Rateau turbine.

Velocity compounding, marine applications

Fig. 5 shows a three-wheel velocity compound turbine arrangement. The steam at inlet pressure expands through a single row of nozzles down to condenser pressure, or to a given back pressure in a non-condensing application. As in the De Laval turbine the interchange of heat to kinetic energy through this expansion results in the steam leaving the nozzle with a large kinetic energy content in the form of a high-speed flow of steam. To overcome the problems this produces in a De Laval type turbine, two or more rows of moving blades are attached to a single wheel, with guide blades between each row. The high-speed flow leaving the nozzle enters the first row of moving blades, where it suffers a change in direction due to the shape of the blades, and as a result of this change of velocity loses some of its kinetic energy to that row, so producing a rotational force. The steam flow then passes across a row of guide blades fixed to the casing, redirecting the steam onto the next row of moving blades to give a turbulence free entry.

On the second row of moving blades, the direction of the steam flow is again altered by the moving blades and more kinetic energy is given up to drive the turbine rotor. The steam is then redirected by a second row of fixed blades onto a third row of moving blades where more kinetic energy is absorbed by the rotor. The steam then passes to exhaust. There is no pressure drop across either the moving or guide blades, the pressure in the casing surrounding these blades being the pressure of the steam leaving the nozzle.

The single wheel carrying the blades may be forged integral with the rotor or keyed or shrunk-force fit onto the rotor. Compared with the De Laval type of turbine, working with the same nozzle pressure drop and hence steam flow speed at the inlet to the blades measured relative to the casing, a velocity compounded turbine gives more acceptable blade speeds, centrifugal stresses and gear ratios for marine applications. This is achieved by dividing the overall change in velocity between a number of moving blades fixed to one wheel, although in practice this leads to some loss of efficiency. It can be seen from Fig. 5 that the speed of the steam flow relative to the casing at the inlet to the rows of moving blades is different at each blade ring, whilst these blades, being attached to the one wheel, have the same mean blade speed, since they are designed with the same mean blade diameter. Therefore the conditions for maximum efficiency, i.e. the blade speed being approximately $0{\cdot}5 \times$ steam speed relative to the casing, and the exhaust steam velocity approximately at right angles to the plane of motion of the blade ring, giving minimum kinetic energy lost to exhaust, cannot be achieved for all the blade

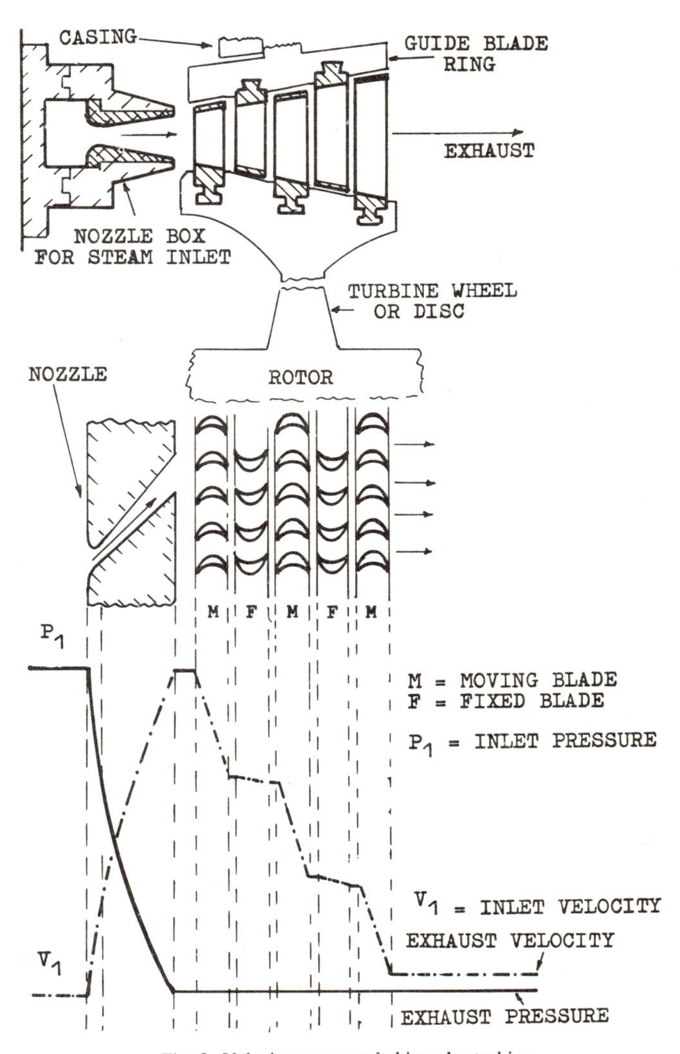

Fig. 5 Velocity compounded impulse turbine

rings at the same time. A design can be produced to improve efficiency by making the blade speed approximately one-sixth of the steam flow speed relative to the casing at the inlet to the first row of moving blades, for a wheel with three rings of moving blades, so that the ideal conditions are approached on the last ring, with the minimum kinetic energy passing to exhaust. For a two row wheel the blade speed should be approximately a quarter × the steam flow speed relative to the casing at the first row of moving blades to attain similar ideal conditions at the final blade row. A further complication arises due to the reduction of the speed of the flow of steam relative to the casing, in that a greater cross-section area is required for the steam flow if the same mass flow per second is to be achieved throughout the turbine. This can be achieved to some extent by increasing blade lengths, but this in turn produces centrifugal stress problems and recourse has to be made to increasing blade exit angles, and hence inlet angles, i.e. flattening the blades. Whilst the former helps to clear steam from the preceding blades, there is a reduction of the energy available to do work on the blades. Again this loss becomes even greater with each successive stage. Also in practice there is an increase in the volume of the steam due to friction re-heating it as it passes across the fixed blades and at the same time reducing the velocity of the steam by some 12 per cent, further lowering efficiency.

Theoretically, for a given nozzle angle and the correct blade speed/steam flow relationship, the efficiency of a velocity compounded turbine is constant whatever the number of rows of moving blades, but in practice the efficiency, and the work done in the last row, drops considerably as the number of rows increases. For a two row wheel, the efficiency is about 68 per cent, and for a three row wheel about 50 per cent. A De Laval type turbine, which may be considered a special case of the velocity compounded design, has an efficiency of about 85 per cent.

The design produces a short, light-weight turbine which is frequently used where space and weight limitations are more important than efficiency, as in astern turbines. The design is also used for feed pump and fan drives as well as at the steam inlet end of HP turbines, where it provides a large initial reduction in steam pressure and temperature, thus shortening and lightening the rotor and reducing the need for high grade alloys for the remaining stages as well as reducing scantling requirements. The nozzle box only is subjected to superheater steam outlet temperatures requiring specialized materials. Gland length and leakage and also blade tip leakage can be reduced. The design also may be known as a Curtis turbine.

Pressure-velocity compounding

A pressure-velocity compounded turbine arrangement is shown in Fig. 6 together with the diagrams showing the behaviour of the absolute steam velocity and steam pressure from steam inlet to exhaust. Part of the overall expansion takes place in the first ring of nozzles, giving an increase in kinetic energy in the form of a high speed flow of steam leaving the nozzles. The velocity drops as the steam gives up its kinetic energy on passing over successive rows of moving blades attached to a wheel or disc which is either forged integral with the rotor or keyed, or a shrunk-force fit onto a central spindle. The exhaust steam from this first stage then passes through nozzles in a diaphragm dividing this from a second similar stage, the steam expanding in the nozzles to exhaust pressure, and gaining in kinetic energy as it does so. This high speed steam flow then passes over a second wheel of similar form to the first one, with the kinetic energy being given up as the velocity drops to the exhaust condition. The design is used in applications where, if the overall expansion was split into a number of small pressure drops to give acceptable blade speed conditions as in pressure compounding, the rotor would be too long, or if velocity compounding was used with the overall pressure drop taking place in one ring of nozzles, the resultant steam flow speed would require unacceptably high blade speeds even for a three row wheel. In theory it would be possible to use a four or five row wheel to give suitable blade speeds but in practice such an arrangement would be very inefficient with very little work being done on the last rows of moving blades. By dividing the overall

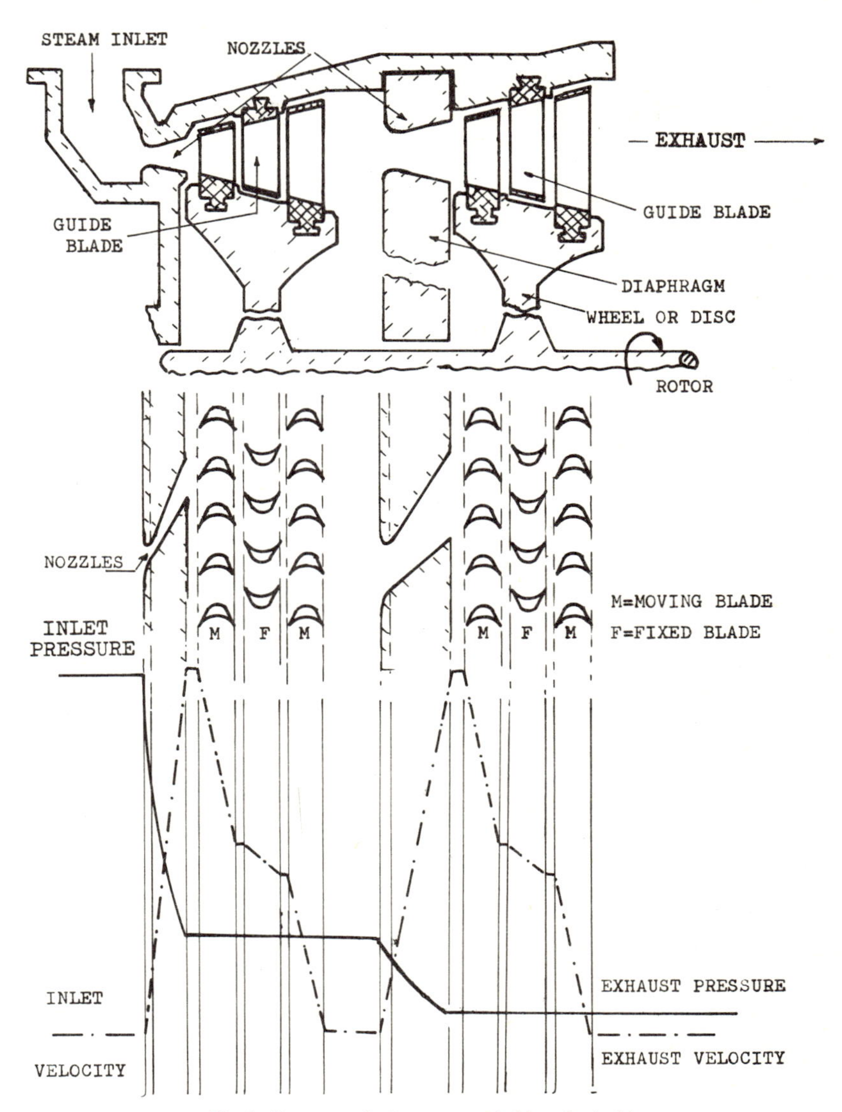

Fig. 6 Pressure-velocity compounded impulse turbine

expansion between two rings of nozzles and using two stages consisting of one wheel with two rows of blades for each stage, steam velocities at exit from the nozzles can be kept at an acceptable level and the blade speeds, centrifugal stresses and gear ratios maintained at a suitable value for marine applications. A typical application would be for an HP astern turbine on the LP turbine rotor, where a short light-weight turbine is required to take a very large pressure drop.

Impulse reaction turbine (Parsons turbine)

Note This has been included for historical interest as this type of turbine is now very rarely found, although the principle of expanding the steam through the moving blades is used in some modern impulse turbine designs to recover some of the energy lost due to friction of the steam as it passes across the nozzles and blades causing a reheating effect.

Fig. 7 shows the behaviour of the absolute steam pressure and velocity in a Parsons reaction turbine. Examination of the diagrams shows that the steam pressure drops not only across the fixed blades or nozzles, but also across the moving blades, this latter fact being the main difference between this and the impulse turbine.

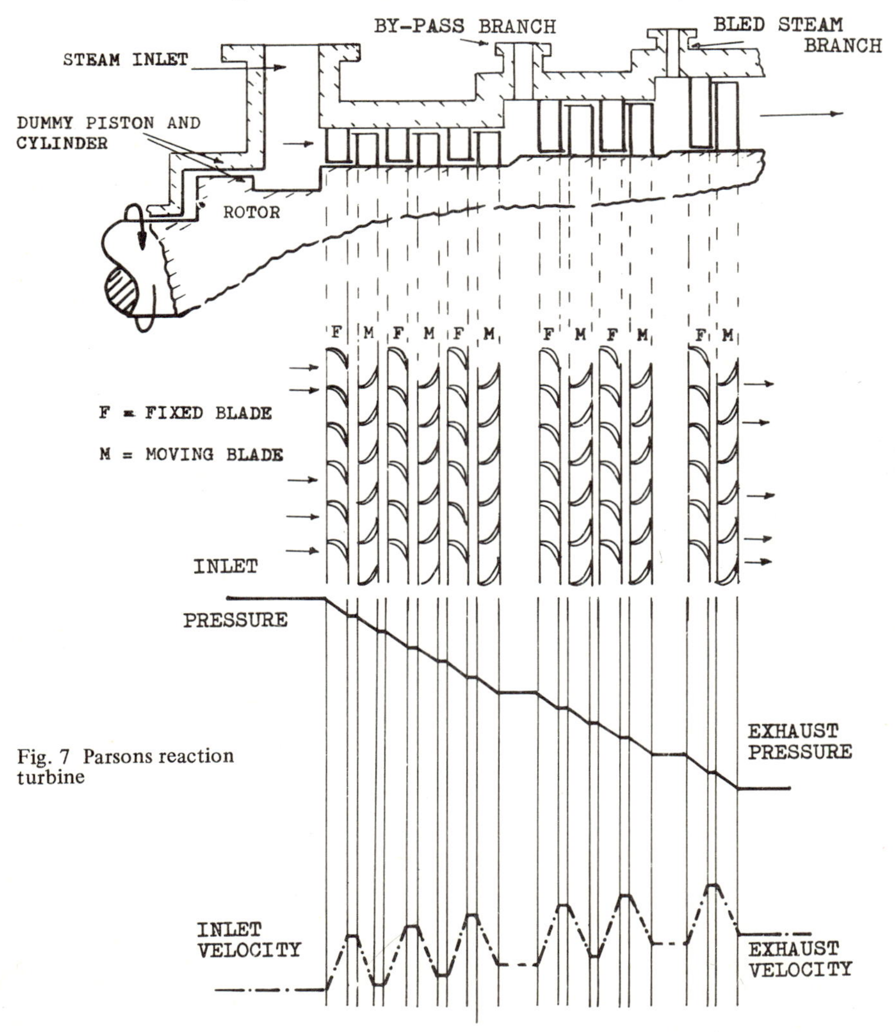

Fig. 7 Parsons reaction turbine

Expansion of the steam in the nozzles formed by the fixed blades increases the kinetic energy of the steam flow so that it passes onto the moving blades at high speed. For maximum efficiency the blade speed should be about $0{\cdot}9 \times$ the steam flow speed measured relative to the casing. The moving blade changes the direction of the steam flow, producing a change in velocity and hence a change in kinetic energy. The rotational energy achieved by this impulse effect is assisted by the expansion of the steam through the nozzle shaped passage between adjacent moving blades. This expansion increases the speed and thus the velocity of the steam flow measured relative to the moving blade so that a further change in energy takes place, imparting an additional rotational force to the blade. This design should therefore more correctly be called an impulse-reaction turbine. If the heat drops produced by the respective expansions across the fixed and moving blades are equal, the design is known as half-degree reaction.

By allowing the overall pressure drop to be split into a large number of small expansions so that increases in steam flow speed were small, blade speeds and rpm could be kept low, and in early designs this enabled the turbine to be coupled direct to the propeller. With the advent of suitable gearing, the expansion over one long rotor could now be divided between two, three or four rotors connected to one shaft through the reduction gears. As steam pressures and temperatures were increased to obtain more power, a greater number of stages were required to deal with the larger expansions and heat drops involved and the consequent increasing rotor length introduced problems with sagging and expansion allowances, together with increased inefficiency due to steam leakage across the blades at the inlet end. Since there is full admission around the first ring of blades and the specific volume of the steam is low, the blade heights required are very small. This feature alone can cause a decrease in nozzle and blade efficiencies due to turbulence and although clearances do increase as the blade length increases, these are not proportional and hence when the blade height is small, the clearance is a larger percentage of it and the ratio of steam leakage to steam doing work is higher. In order to keep these losses as small as possible, fine clearances are essential and although under steady steaming conditions such clearances are feasible, when manoeuvring etc., larger clearances are necessary to prevent contact between rotor and casing components due to the fluctuations in differential expansions that take place under the influence of superheat steam temperature variations. 'End-tightened blading' is frequently used on high pressure impulse-reaction turbine rotors to accommodate these conflicting requirements. Also, to keep this annular leakage as small as possible, these rotors tend to have a smaller diameter at inlet than impulse turbines. The mass flow of steam per unit time through the turbine must be kept constant, but as the pressure drops across each row of blades, the specific volume increases, requiring an increase in axial velocity or greater blade cross-section area for the steam to flow through, or both. If the blade angle is altered it means a new set of blades of different dimensions for each ring, whilst if the height is adjusted, as well as altering the blade speeds, the casing or rotor would have to be stepped at each ring to accommodate this change. Each method involves considerable extra manufacturing costs and therefore a compromise is adopted by dividing the turbine into a number of sections and making the fixed and moving blade shapes identical in each of these. The section would be designed to accommodate the particular requirements of the steam expansion at mid-point. The casing and/or rotor is also stepped to suit the blade requirements at each stage. Blade heights can only be increased a certain amount relative to the rotor diameter due to centrifugal stress and steam flow problems. Once the blade height exceeds approximately 0·2 × mean diameter, the latter has to be increased. Hence the rotor is sometimes stepped, particularly towards the LP exhaust stages, where the slope at each stage may be extremely pronounced.

High-pressure turbines of this type usually incorporated end-tightened blading, as mentioned previously. Both fixed and moving blades of this type incorporated shrouding around the tips with the edge thinned. The rotor incorporated an adjusting arrangement which allowed the rotor to be moved axially for'd and aft, so that for normal running conditions the shroud/root clearance was about 0.254mm while for manoeuvring conditions it was increased to about 1.254mm.

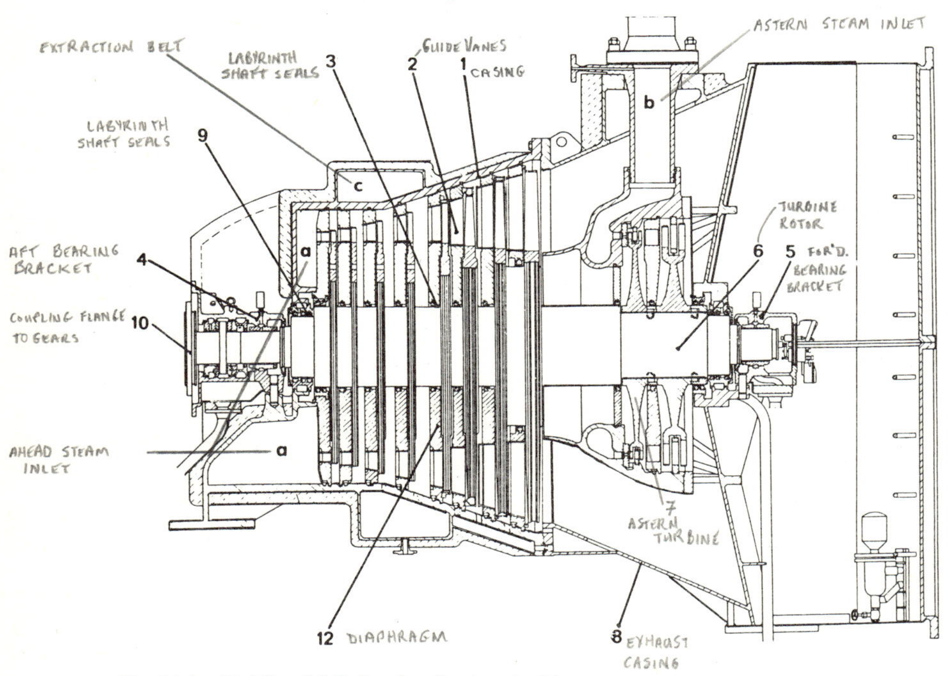

Fig. 8(a) Stal Laval LP ahead and astern turbines

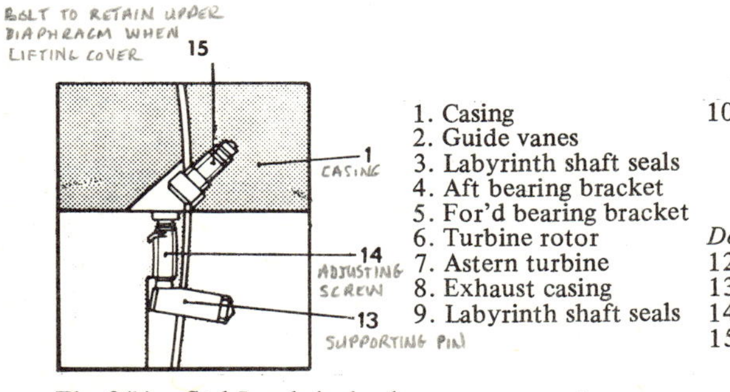

1. Casing
2. Guide vanes
3. Labyrinth shaft seals
4. Aft bearing bracket
5. For'd bearing bracket
6. Turbine rotor
7. Astern turbine
8. Exhaust casing
9. Labyrinth shaft seals
10. Coupling flange to gears
    a. Ahead steam inlet from HP exhaust via crossover pipe
    b. Astern steam inlet
    c. Extraction belt

Detail

12. Diaphragm
13. Support pin
14. Adjusting screw
15. Bolt to retain upper diaphragm when lifting cover

Fig. 8(b) Stal Laval single plane arrangement

Single plane turbine

This arrangement is based on a cross-compound two-cylinder design (steam passes through the ahead HP turbine and then exhausts into a separate LP turbine where it expands down to the condenser conditions) in which the LP turbine has an axial exhaust to the condenser through a diffuser formed naturally by the outer casing and the casing around the astern turbine. The latter also exhausts in the same direction towards the condenser which is on the same plane as the LP turbine. Thus the two exhausts have the same direction of flow as against opposing directions of flow involved in an underslung condenser configuration. The axial flow arrangement results in a 50 to 60 per cent reduction in exhaust losses (1 per cent gain in fuel economy) compared with an underslung design, while the condenser has a cylindrical cross-section giving improved flow properties and lower manufacturing costs compared with an underslung condenser which is frequently made rectangular in cross-section in order to allow it to be positioned between the LP turbine and the tank top. The saving in engine height has made it possible to put the boiler above the turbines allowing a shorter engine room to be achieved (Figs 8a, b, c).

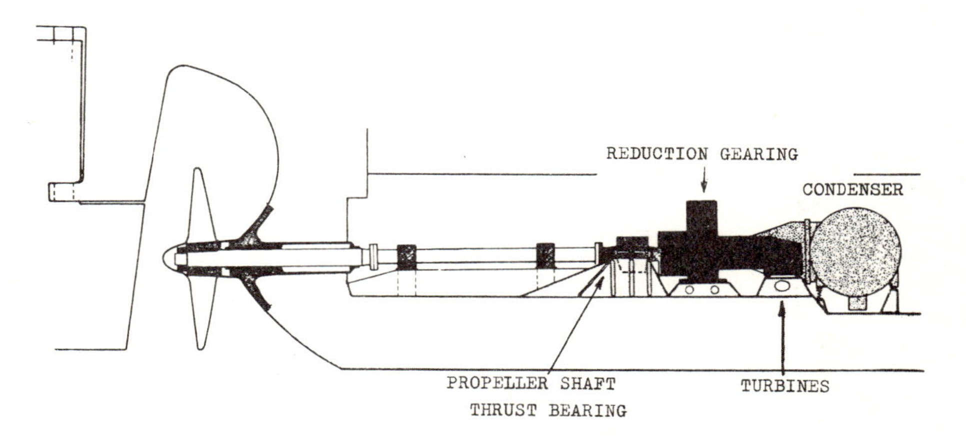

Fig. 8(c) Single plane layout

CHAPTER 2

Turbine Construction

Gashed disc rotor, materials and tests

In most cases modern HP turbine rotors are of the Rateau or pressure compounded design employing only a few stages for the expansion of the steam, about eight to ten, to give the required heat drop. This arrangement produces a shorter rotor and provides savings in weight and overall length. Such rotors are usually solid forged, providing a homogeneous rotor with an even grain flow, giving an even expansion, good thermal stability with less likelihood of distortion under high temperatures.

After forging, the rotor is machined to produce the necessary discs or wheels to carry the blades, the wheels being frequently made of uniform thickness, although where stresses demand it, the disc thickness may be slightly increased near the shaft. This form of construction is also used for the LP turbine rotors, which may have from seven to nine ahead stages and two or three astern stages. After rough machining the rotor is subjected to a thermal stability test and following final machining and the fitting of the blades it is subjected to static and dynamic balancing tests. The rotors of this form of construction generally have a first critical speed, a speed at which very heavy vibration may occur in a turbine that normally runs smoothly, well below the normal running speed. This critical speed is usually indicated on a plate near the controls and under no circumstances should the turbine be operated continuously at this speed.

The design, as shown in Fig. 9, is known as a gashed disc rotor and one feature is that it allows the minimum rotor diameter to be used where it passes through the diaphragms so that the area for steam leakage here is kept as small as possible. Holes

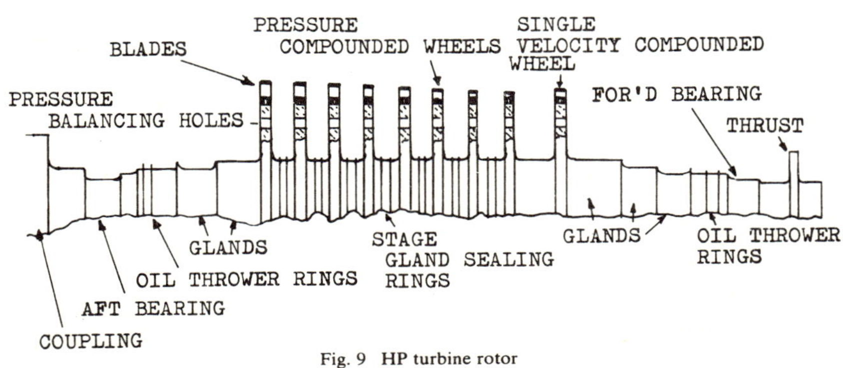

Fig. 9 HP turbine rotor

are usually made in each disc to ensure that there is no build-up in pressure on one side to create an end thrust. For a rotor operating up to 566°C, a typical material specification is as follows:
0·27 to 0·37% Carbon
0·70 to 1·00% Manganese
0·040% Phosphorus
0·040% Sulphur
0·15 to 0·35% Silicon
0·85 to 1·25% Chromium
0·50% Nickel maximum
1·00 to 1·50% Molybdenum
0·20 to 0·30% Vanadium

LP turbine rotor. Precautions for warming-through and manoeuvring. Use of Guardian valves.

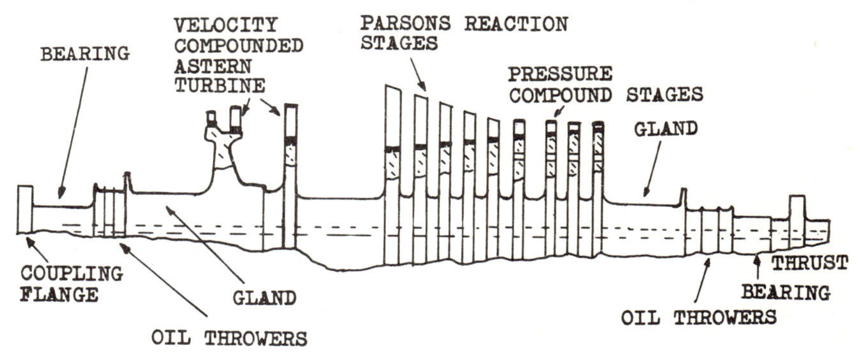

Fig. 10 LP turbine rotor

Fig. 10 shows a typical gashed disc LP turbine rotor operating at about 3700rpm. The ahead turbine has four rows of impulse blading followed by five rows of Parsons reaction blading, although the impulse blading may have up to 20 per cent reaction effect at the mean blade height. The astern turbine consists of a single wheel two stage velocity compounded turbine followed by a single stage velocity compounded turbine. For this type of turbine the steel used for the rotor would have:

0.3% Carbon
0.25% Silicon
0.6% Manganese
2.5% Chromium
0.4% Nickel
0.45% Molybdenum
0.03% max. each Phosphorus and Sulphur

The blades would be of stainless iron with:
0·04% Phosphorus maximum
11·5 to 13·5% Chromium
1·0% Nickel maximum
0·12% Carbon maximum
1·0% Manganese maximum
1·0% Silicon maximum
0·04% Sulphur maximum

The LP turbine rotor is very susceptible to distortion when warming through and when manoeuvring due to the comparatively low temperature of the condenser on the underside tending to cause this side to contract compared with the upper side. The upper side is, however, subject to the higher temperature of the warming through steam as it tends to rise and is liable to a slight expansion so that if the rotor is not turned regularly, there is a danger of distortion. On older designs particularly, the long distance between the bearings, due to the large number of reaction ahead stages, dummy pistons and astern turbines, can aggravate the distortion considerably. Such a distortion can cause very serious vibration and may bring the rotor into contact with the casing causing a rub and the possibility of severe damage to blading or glands or a bent rotor. When warming through, the LP rotor should be continuously rotated, but if this is not possible, then turned at least every ten to fifteen minutes ensuring that the rotor finishes in a different position each time. The vacuum should be kept as low as possible, just sufficient to create a steam flow. When manoeuvring, if the turbines are stopped for any length of time, the vacuum should be dropped back and the LP turbine rotated frequently.

When manoeuvring, LP turbines are subject to considerable fluctuations in temperature, particularly when going astern. Astern turbines are usually kept as small and as light as possible to reduce the length of the turbine and the weight. They are not particularly efficient and therefore the exhaust steam tends to be hot and the casing can reach a high temperature if astern running is required for a comparatively long period of time. This may also cause both rotor and casing distortion and the maximum temperature for the casing given by the manufacturer should never be exceeded. In LP turbine designs with underslung or down-flow condensers, the ahead and astern turbines exhausts face one another and despite the fact that deflector plates are fitted to both to induce the steam to flow down into the condenser, some steam is picked up by the astern turbine when running in the ahead direction and vice versa. The turbine picking up the steam is running in the wrong direction with regard to the steam flow and tends to compress and churn the steam—an effect known as windage—which can cause the casing and the turbine rotor to overheat and distort. The danger can be seriously aggravated if the LP manoeuvring valve leaks when the turbine is running ahead, and to reduce the likelihood of this happening, a double shut-off valve or Guardian valve is fitted between the manoeuvring valve and the turbine, with a drain connection between the two valves. This may be led to the condenser and have a test cock or pressure gauge fitted to check for any signs of leakage.

With single plane turbines and other designs with smaller LP rotors than were used in the past, it will not be necessary to reduce the vacuum when on Stand-by or stopped for any length of time when manoeuvring as these are less susceptible to distortion but there must be continuous or frequent rotation to prevent trouble.

Built-up rotors

A common form of construction used for impulse turbines operating at moderate temperatures and rpm and for LP reaction turbines is the built up rotor as shown in Fig. 11. It consists of a number of discs prepared from a circular ingot of steel, forged or pressed into disc form. These are then machined and fitted to a shaft which may be of stepped or uniform diameter according to the method employed for attaching the discs. These are hydraulically pressed on to the shaft when hot to give a combined shrink and forced fit. The difference in diameters must be closely watched so that even when the wheel is running at overspeed (usually about 10 per cent above the operating speed) there is some mutual pressure between the wheel and shaft. When starting a turbine from cold, the discs will reach the working temperature more quickly than the shaft and unless there is a sufficient force fit allowance, they may become loose. The torque is transmitted by sunken keys, one key for moderate torques, with the keys on successive wheels being arranged at 180° to each other. For greater torques two or more keys may be used. A small axial clearance is left between adjacent discs to allow for expansion due to temperature differences, the shaft being stepped to allow each wheel to pass over the shaft until it reaches its own seat. To prevent any appreciable difference in steam pressure being set up across the two sides of the disc, pressure balancing holes are drilled from one side to the other.

The objection to forcing the wheels onto the shaft is that the stresses in the wheel are increased due to the heavy force fit, and under high temperatures the wheel may be subject to creep. The effect of this is that a turbine wheel rotating in high temperature steam conditions and subject to radial and tangential stress may gradually change in size, the diameter increasing at a slow but steady rate. Such a change in dimensions due to creep reduces the pressure due to the force fit and the wheel tends to loosen. Fretting corrosion may also occur where the disc and shaft are in contact. This form of construction is rarely, if ever, used except for astern turbine disc attachment to rotors by some manufacturers.

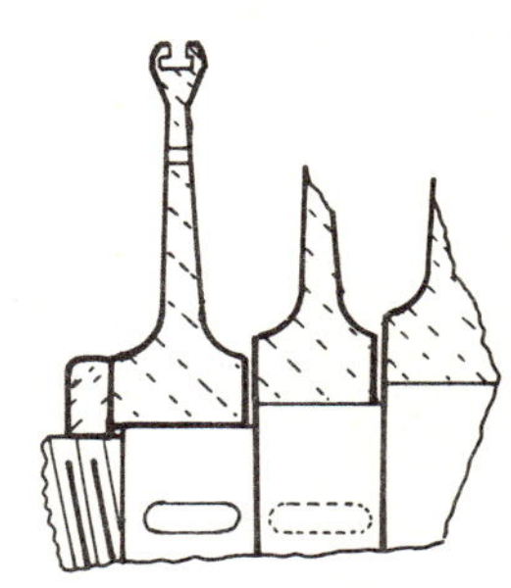
Fig. 11 Built up rotor

Double casing turbines and double flow turbines. Materials.

Fig. 12 shows a typical LP double casing and double flow turbine. A double casing turbine consists of two casings, one inside the other and the arrangement may be found in use on HP and LP turbines as well as on single cylinder turbines. Steam is supplied to the inner casing where it expands through the turbine and exhausts into the space between the two casings, so that it surrounds the inner casing whilst the outer casing is subject to exhaust pressure. In this way the rotor journal glands are only subject to exhaust pressure, and in the case of an HP turbine these glands operate at far less arduous conditions than in a single casing design. The pressure drop across the HP inlet gland is far less, whilst the temperature gradient across the casing is considerably reduced with less likelihood of distortion occuring. With all types of turbine the outer casing is at lower temperature and much less heat is given up to the engine-room, whilst, due to the complete jacketing of the inner casing with steam, warming through can be carried out more quickly. The expansion and contraction of crossover pipes between the turbines is absorbed by the outer casing so that side thrusts on the turbines are greatly reduced making misalignment between the rotors and gear pinions less likely.

Double flow turbines are used mainly for LP reaction turbines and the design produces an alternative means of balancing the end thrust produced in this type of turbine to the dummy piston and cylinder arrangement. The design allows the steam flow to be split, half flowing for'd and the other aft, so that the last few stages of the turbine can use shorter blades. This reduces the problem of centrifugal stresses on the blades and allows smaller blade angles to be used. However, the design does produce a long rotor which tends to be very susceptible to sagging and distortion. The design incorporates a two stage velocity compounded astern turbine and the hood around the lower section of the first row of moving blades of this turbine should be noted. This arrangement is commonly used with this type of turbine where partial admission is employed to reduce the windage, pumping losses and dangers of overheating that can arise when the blades churn up the steam carried round from the section where steam is admitted.

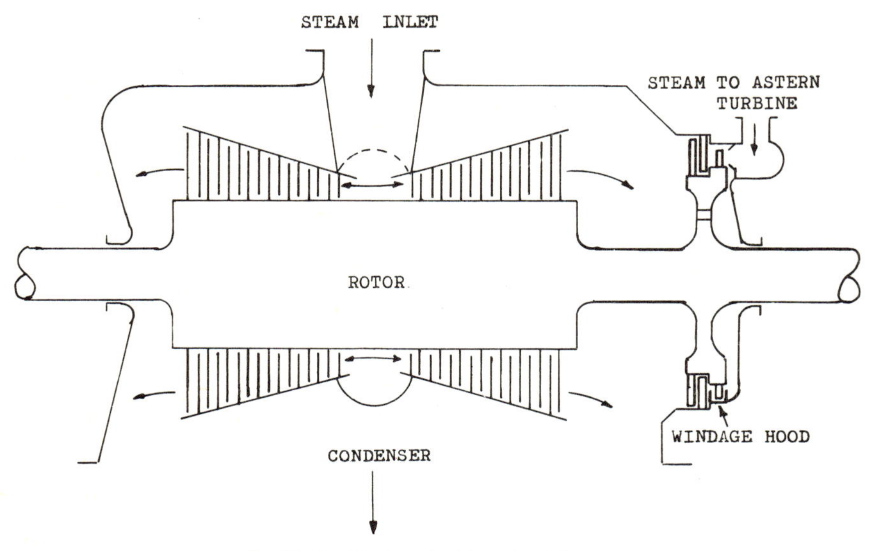

Fig. 12 Double flow, double casing, LP turbine

A material quoted for turbine casings operating in the 427° to 538°C temperature range comprises: 0.25% Carbon (maximum)
0.8% Manganese (maximum)
0.6% Silicon (maximum)
1.0–1.3% Molybdenum
0.25 to 0.35% Vanadium
1.2–1.6% Chromium

Such double flow turbines will be rarely, if ever, found in modern practice.

With castings, the all-important feature of castability exerts a strong influence on design. It often dictates a wall thickness two or three times thicker than that necessary to withstand surface conditions if a sound casting is to be ensured. For low-pressure turbine casings with temperatures up to 360°C cast iron may be used. Up to 400°C plain carbon steel castings are used. Above 400°C low-carbon steel incorporating molybdenum and chromium, or chromium and vanadium, may be used to provide superior creep resistance. The materials should be suitable for welding so that turbine casings with complicated shapes need not be made from a single casting. Casing stud materials:

0.20 – 0.25% Carbon
0.20 – 0.40% Silicon
0.40 – 0.70% Manganese
11.5 – 12.5% Chromium
0.8% Nickel (maximum)

1.0 – 1.4% Molybdenum
0.03 – 0.06% Niobium
0.25 – 0.35% Vanadium

Tensile strength 980 N/mm²

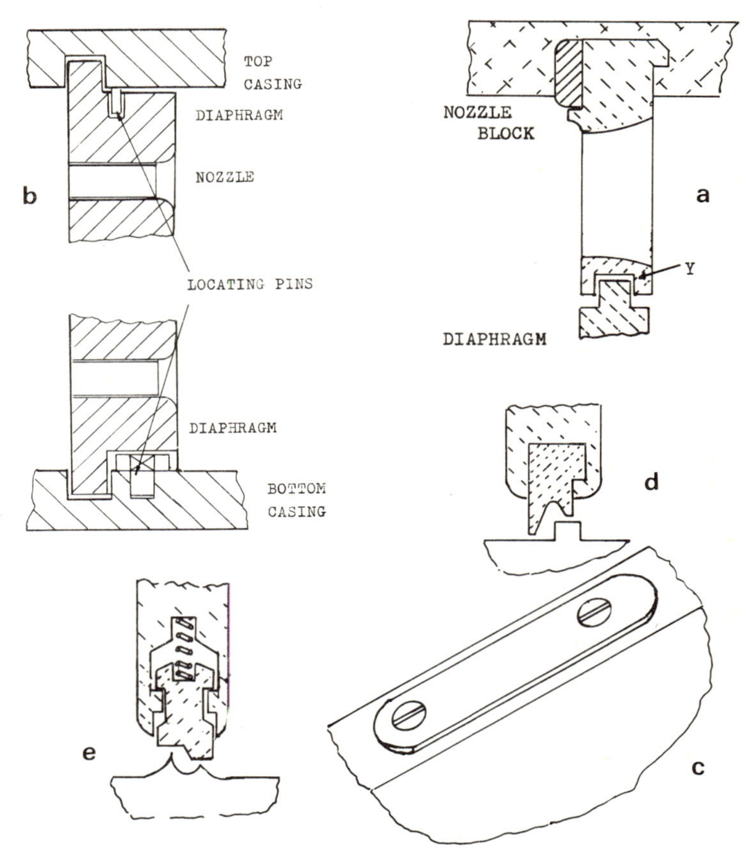

Fig. 13 Diaphragm expansion and glands

Fixing and expansion arrangements for diaphragms. Diaphragm glands.

Diaphragms are used in impulse turbines to divide the overall expansion into the requisite number of stages. As such they present complex structures with large surfaces in contact with the surrounding steam in relation to the mass of the various components, and yet requiring sufficient strength and stiffness to withstand the pressures drops existing in the turbine without being excessively wide. The steam in contact with the surfaces of the diaphragms may be at different temperatures, and they are liable to very rapid expansion and contraction as steam temperatures change during speed alterations. At the same time each one behaves in a different manner due to variations in sizes so that, whilst freedom of expansion is necessary, close alignment is essential where the shaft passes through to prevent excessive steam leakage. The axial width of the diaphragm is important as this contributes a significant amount to the width of the stage and thus the number of stages that can be accommodated over a given bearing centre distance. The critical speed of the rotor then becomes involved and then the number of turbines that may have to be used to deal with a specific expansion requirement.

Fig. 13 shows two methods employed for securing diaphragms in a turbine casing and yet allowing for adequate expansion. In diagram (a) the nozzle block is fixed firmly in the casing by means of a caulking piece, whilst expansion is allowed for between the diaphragm and the nozzle block at point Y, the distance being about 2·4mm. Radial keys help to position the diaphragm relative to the nozzle block.

Diagram (b) shows an arrangement where the diaphragm and nozzles are welded together to form a solid unit, and expansion has to take place at the casing. The latter is recessed and a projection around the diaphragm fits into this recess. Locating pins are fitted on the vertical centreline of the top and bottom sections. See also Fig. 8.

At the horizontal joint between the two diaphragms a key in one half fits into a recess in the other, holding them in alignment and preventing steam leakage from one side to the other, diagram (c). When steam is admitted to the turbine the pressure drop across the diaphragm holds it hard against the down-stream face of the groove in which it is fitted in the casing, preventing steam leakage around the ends of the diaphragm. Steam leakage where the rotor passes through the diaphragm at the centre is controlled by labyrinth packing as shown in diagrams (d) and (e). Steam from the high pressure side passes through the restricted annular space between the rotor and the packing and is throttled in the process, the increase in velocity resulting from this being dissipated in the enlarged space between the two restrictions in the form of turbulence, and conditions allowing almost complete throttling of the steam from the initial pressure to the pressure in the space into which the packing discharges can be achieved, thus limiting the steam flow. In some designs the packing is caulked into the diaphragm as shown in diagram (d), although in more modern temperature turbines spring-loaded packing is used as shown in diagram (e). Heat resisting radial steel springs hold the packing against a shoulder in the diaphragm, giving the required radial clearance between the rotor and packing, usually about 0.5mm. This design has the advantage that should contact occur between the rotor and the gland due to some form of distortion, the spring allows the gland to give and prevents a heavy rub occurring. Should a rub occur with the gland as shown in diagram (d), considerable heat can be generated between the rotor and the gland, causing the rotor metal to expand, increasing the pressure until the rotor becomes red hot in that particular spot and warps. This can cause a permanent bend in the rotor, which would then have to be removed for straightening. Diaphragm material is of 0.18% C, 0.2% Si, 0.65% Mn, 1.0% Cr, 0.4% Ni, 0.6% Mo. Gland ring material is 0.08% C, 1.0% Si 1.0% Mn, 13.0% Cr, 0.5% Ni.

In some cases for low temperature applications, cast iron may be used for the construction of the diaphragms, and flat springs may be used in place of the radial type for the spring-backed glands. Pressure drops across the diaphragms may be in the region of 207 to 280 meganewtons per square metre, and such loadings may produce deflections in the region of 2mm.

In many modern designs the diaphragms are of welded construction, the nozzle plates or guide vanes being inserted into punched slots in an inner and outer steel ring. The whole assembly is then welded onto the centre body and a peripheral supporting ring as shown in Fig. 14(a).

Blind holes or recesses are usually made in the sides of the diaphragm in which lifting clamps can be placed to facilitate removal from the casing. Nozzle

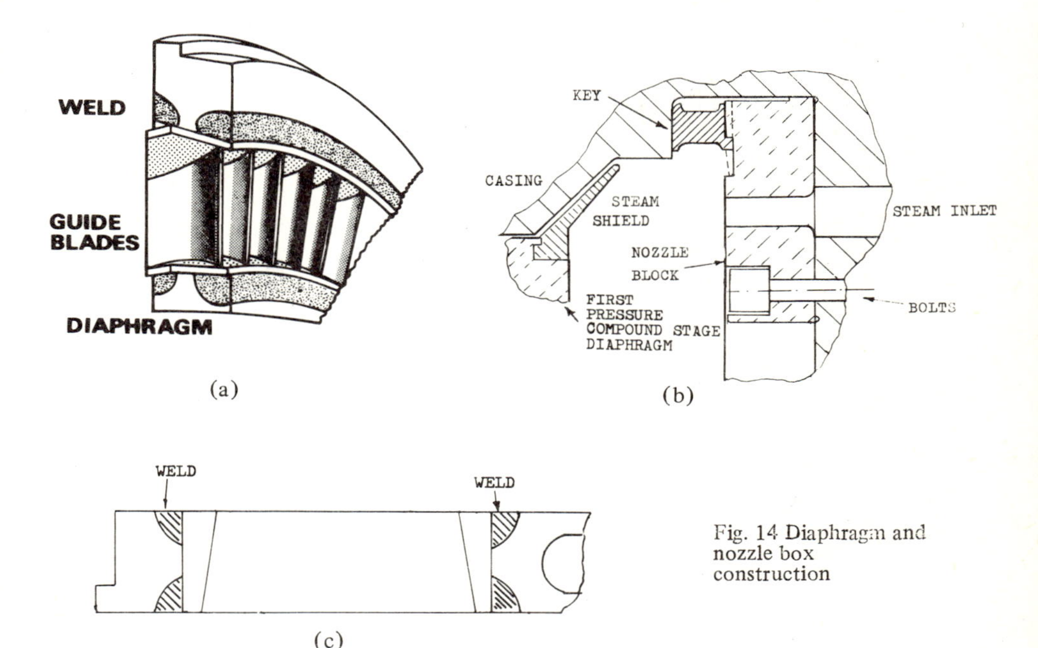

Fig. 14 Diaphragm and nozzle box construction

blades or guide vanes are commonly made from stainless iron:

0.09% Carbon
0.22% Silicon
0.4% Manganese
12.0% Chromium
0.6% Nickel
0.5% Molybdenum
0.03% Sulphur
0.02% Phosphorus.

Fig. 14(b) shows the method of construction used at the HP inlet. Depending on the particular design, the nozzles may provide a full 360° arc admission but usually partial arc admission is employed. The inlet nozzles are split into groups, each group being controlled by a separate valve and only a given number of nozzles are used to allow the maximum expansion ratio for the steam for the power required. Fig. 14(c) is a simplified version of the welded design.

Fig. 15 shows a steam inlet system. Here steam from the boilers is supplied to the manoeuvring valves (1) and (2) via two supply lines (a). An isolating valve (4) is fitted to each supply line and in order to facilitate the opening of these a pressure balancing valve (5) is fitted in line (b) which should be opened first. The supply lines are drained through (c) while the valve casings are drained through (e) to the high-pressure drains tank, and any leakage from the manoeuvring valve glands passes to the gland condenser through line (d).

For warming-through the manoeuvring valves are shut and sufficient steam for this purpose is supplied through valve (6).

For Ahead operation the manoeuvring valve (1) is opened and steam is supplied through nozzle group (7A), while the astern manoeuvring valve and

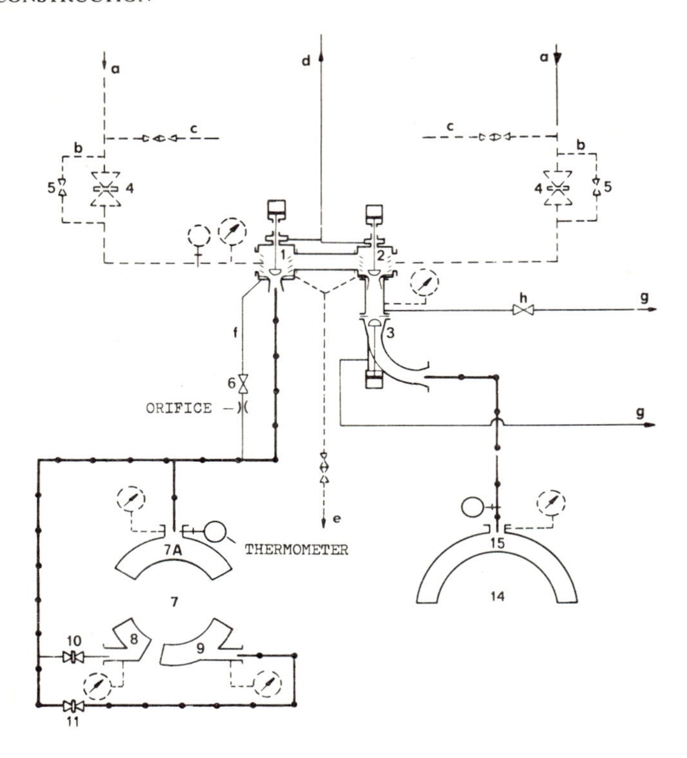

Fig. 15 Stal Laval steam inlet system

Guardian valve are obviously shut (valves 2 and 3 respectively). The space between these valves is connected to the condenser via valve (h) which has a drilled head to provide continuous drainage. When the steam requirement is beyond the capabilities of the nozzles in (7A), nozzles in groups (8) and (9) can be brought into use as required via valves (10) and (11).

For Astern operation the astern turbine (14) is supplied with steam via valve (2) and the Guardian valve (3) to the nozzle box (15) supplying steam to the astern turbine. The ahead valves are then obviously shut.

In the diagram (f) is the warming-through line, (7) represents the ahead turbine and (g) is a drain to the main condenser.

Shrouding, lacing wires and damping wires

As the blades of a turbine pass across the nozzles in the diaphragms, the jets of steam issuing from the nozzles in rapid succession can create pulsating forces on the blades and set up vibrations which can result in cracks developing in the blades and their subsequent failure. Variations in the steam flow due to partial admission in the HP turbine inlet stages and due to the extraction of steam at bleed points can also have a similar effect. The vibrations produced are complex and can result in a bending action, or a twisting action, or a combination of both with one or more node points along their length (Fig. 16). Long thin blades similar to those used for LP turbines may not be very substantial and may spread or become incorrectly spaced at the tips. It is common practice therefore to support the blades in some way to prevent this vibration and spreading, usually by the use of lacing wires or shrouding, or a combination of both.

Centrifugal force acting on the steam tends to make it crowd towards the tips, while when the steam makes a sharp turn in the blade passage at high speed it tends to crowd towards the concave face. Both effects tend to make the steam spill over the tips if they are open. Where there is some reaction effect present with a pressure drop across the moving blades, having them open at the tips allows the steam to leak through the radial clearances without performing work on the blades. Where the steam has a low specific volume in the higher pressure stages and the radial clearances are large compared with the blade length, this loss can be serious. There, in addition to supporting the blades **shrouding** is used to prevent steam spillage.

The blades even when fitted with shrouding are subject to damaging vibrations which can cause not only the blades to bend but also the shrouding, and this can result in the shrouding loosening or breaking as well as the blades.

In LP turbines, where shrouding could adversely affect the drainage of water from the blade tips, the blades are left open and radial clearances employed. As the specific volume of the steam is high and the clearances are small compared to the blade length, steam leakage losses are comparatively low and water clearance is of greater importance. **Lacing wires** are used here, one or two rows being employed depending upon the blade length and the amount of support required.

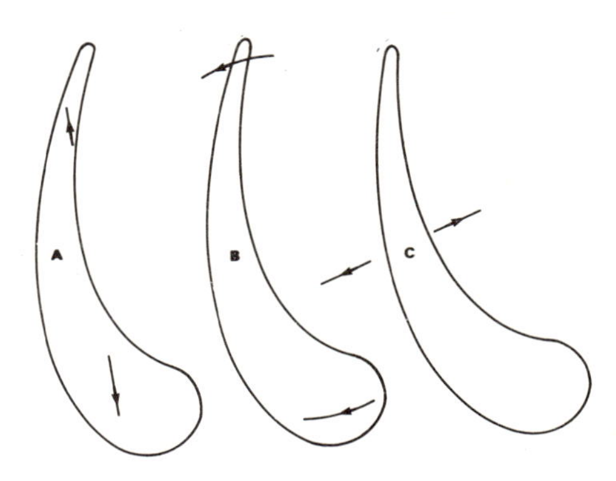

Fig. 16 Blade vibration modes: (A) edgewise, (B) torsional, (C) flapwise

With radial clearance blades, the tips are usually thinned down so that if contact occurs between the tip and the casing the thin section rapidly wears, limiting the amount of damage.

Shrouding may be fitted to the blade tips by riveting, brazing or welding over a tenon as in Fig. 17(a), or riveting (b) where the method used previously may produce cracks under the tenon head or distort the blade. Here the shrouding is machined integral with the blade, the blades being fixed together by a strip of metal passed through the top of the shrouding and riveted to each blade. In other cases the shrouding is again machined integral with the blade and the sections are secured together by fitting a wire through an open groove in the top of the shrouding and then rolling the top edges of this groove over the top of the wire, as in (c).

Lacing wires may be fitted through holes drilled in the blades and brazed in position, or the wires may be in the form of a tube and crimped either side of the hole. In some cases, to assist steam flow the wires may be of aerofoil section. Expansion gaps are usually allowed in both shrouding and lacing wires.

Damping wires are used to assist in reducing vibrations by producing a friction effect on the blades. The blade material itself, and the steam atmosphere, will help to reduce the effect of vibrations while the damping wires, which consist of a loose wire running through holes in the blades, will produce an added vibration reduction effect by limiting the amplitude of vibration.

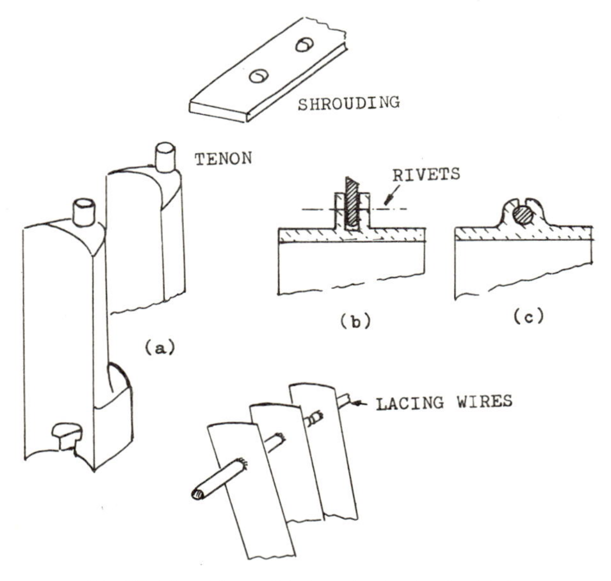

Fig. 17 Shrouding and lacing wires for turbine blades

Blade fixing arrangements

The high rotational speeds of modern turbine rotors produce severe centrifugal stresses which require very often complex root fixing arrangements for the blades to provide adequate resistance. Fig. 18(a) shows a 'T' straddle-root type of blade suitable for impulse turbines together with a suitable gate. At one or more points around the disc the T-section is cut away to allow the blades to be fitted. When the wheel has been completely bladed, the last blade is fitted over this plain section and riveted into position. The blade as shown in (b) is fitted axially and crimped at either side into the root to prevent displacement. The root form shown in (c) is used on blades subject to high stresses, and is designed so that additional contact surface can be gained without weakening to any great extent either the blade root or wheel rim sides. This could occur if the blade root or wheel rim sides are cut deep enough to provide adequate surface area to withstand the load. This 'fir tree' root design can be fitted axially as with the

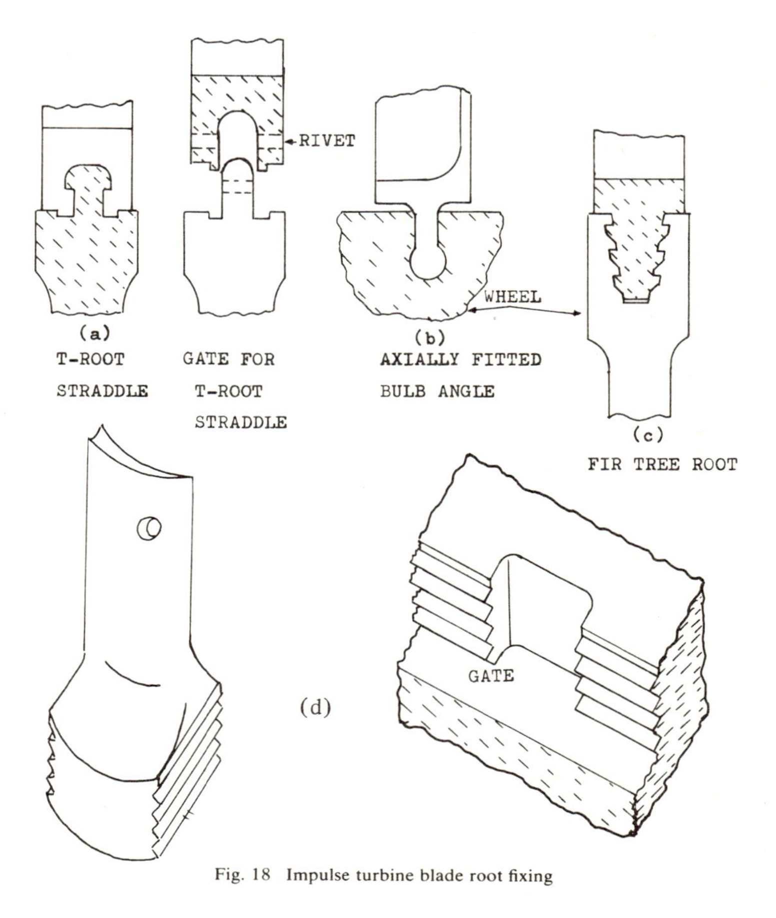

Fig. 18 Impulse turbine blade root fixing

bulb angle or fitted into a groove cut in the periphery of the wheel. In the latter case a gate has to be cut into the periphery of the wheel to allow the blade to be entered. Once entered into the groove the blade is tapped round until it comes against a stopper adjacent to the other side of the gate. Once the groove is full the stopper is removed, the last blade fitted and a stop piece fitted into the gate. This is usually secured with a countersunk screw and the edge peened over. The gates are usually staggered around successive grooves along the length of the rotor to assist balancing. Fig. 18(d) shows a typical blade and gate arrangement.

A material suitable for turbine blades operating over the steam temperature range 400° to 560° would have the following constituents:

0.2–0.25% Carbon	1.0–1.4% Molybdenum
0.2–0.4% Silicon	0.25–0.35% Vanadium
0.4–0.7% Manganese	0.035% max. each Phosphorus and Sulphur
11.5–12.5% Chromium	0.03–0.06% Niobium
0.8% max. Nickel	

Casing blades. Blade stresses and defects

Being stationary the casing blades of a Curtis turbine are not subject to centrifugal stresses and therefore they do not require a complicated root form. They

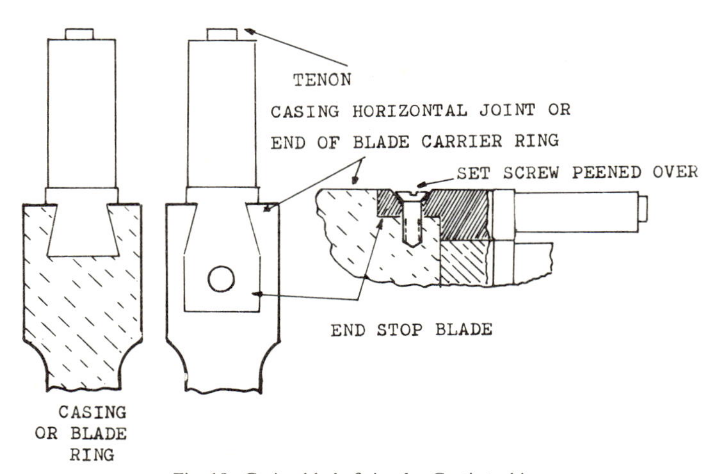

Fig. 19 Casing blade fixing for Curtis turbine

should, however, be a tight fit in the groove as they can vibrate, which in turn may cause fretting. Frequently the design of the turbine calls for the nozzles and guide blades to cover only a comparatively small inlet arc and a carrier ring may then be fitted round part of the casing to carry the blades. End stop blades are required at the horizontal joint or at the ends of the blade ring to keep the blades in position—a typical arrangement is shown in Fig. 19. The predominating stress on a turbine rotor blade is centrifugal stress which is radial in its action and produces a tensile stress over the root section. The force exerted by the steam driving the turbine is tangential

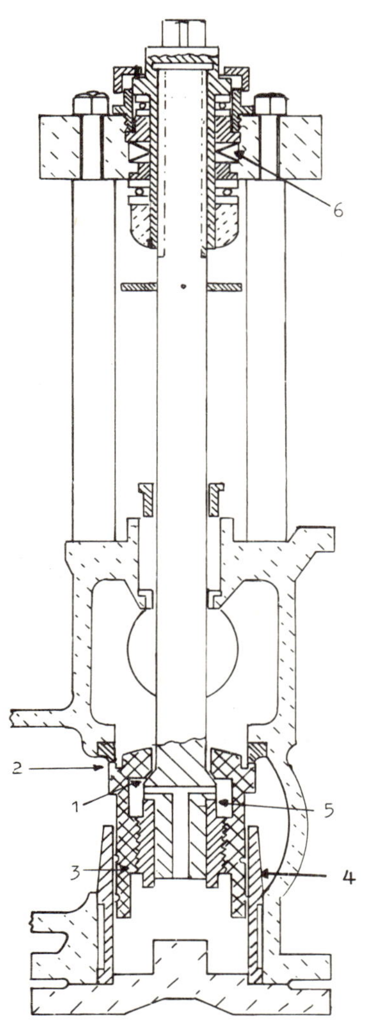

Fig. 20. Manoeuvring valve

and acts at near mid-height of the blade producing a bending stress which is largest nearest the blade root. Variations in steam flow through the nozzle and partial admission can cause fluctuations in this load which set up vibrations in the blade which can eventually lead to cracking at the root and failure. This stress can be very large when starting or reversing. Sudden changes in load on an alternator can also cause speed variations which can add to stresses already present. The blades may also be subject to corrosion from chemicals carried over with the steam from the boiler and this carry-over may also cause deposits on the blades obstructing steam flow and increasing friction. On HP plants silica may cause such a problem particularly at turbine temperatures of about 110°C where it forms a hard coffee-coloured deposit on the blades. Deficiencies in the treatment of the boiler water eventually cause deposits in the turbines which change the steam flow areas resulting in pressure changes. If the deposits occur in the nozzle area the steam flow will be reduced, with pressures dropping throughout the turbine and the output falling.

If the flow area of other stages is reduced the steam flow will remain fairly constant, but stage pressures in the turbine will adjust to allow for a higher pressure drop over each of the affected stages. For example, if the trouble occurs after the first stage, the pressure before that stage will increase, the first, second and third bleed pressures will fall, while the cross-over pressure will initially fall, then steady. In order to obtain a reasonably accurate test of the plant performance the bleeds should be shut off so that the turbine is isolated and there is no interference from the rest of the steam cycle. Excessive end thrust on the rotor may also arise.

Manoeuvring valve

Fig. 20 shows a typical manoeuvring valve suitable for a high pressure, high temperature steam turbine plant.

The valve, in common with many other designs, incorporates a supplementary valve (1) and a main valve (2), with the steam pressure responsible for holding the main valve shut when the valve is in the closed position. In this situation steam leaks up through the 0.254mm radial clearance between the main valve piston (3) and the liner (4). Having no escape path, this steam builds up pressure to that of the superheater outlet, on the main valve, pressing the valve lid onto its seat with a load of approximately 70kN.

When the valve is opened, the initial movement of the hand-wheel lifts the valve spindle until the collars come against guide (5). This initial movement of approximately 0·5mm allows the steam trapped in the space above the main valve to pass down a hole drilled into the spindle, past the supplementary valve seat (1) to the turbines. This allows the main valve lid to become balanced and very little effort should now be required to open the main valve admitting steam to the turbines. When the valve is closed the reverse happens, the initial movement of the spindle moves it down so that the supplementary valve is closed, and the main valve may then be closed easily, the pressure building up on the back of the main valve as this valve closes, the steam pressure eventually holding the valve shut as described previously.

The valve spindle and operating sleeve run in ball races at the hand-wheel end and it is possible to feel the supplementary valve shut as the valve is closed. The hand-wheel can be moved a further half-turn after the valve is fully closed, this movement tending to compress the belleville spring washers (6). If the hand-wheel is turned too far in the shut direction, all the free travel in the belleville springs can be taken up, and the mechanism may then jam due to differential contraction producing very high stresses. Care should be taken not to jam the manoeuvring valve thus and any differential expansion may then be easily absorbed by the belleville washers. Under these conditions the valve should be steam tight and attempts to reduce any leakage by jamming the valve shut are useless considering the load being exerted by the steam pressure on the valve lid. In such conditions the valve should be overhauled. The valve lid, bush and spindle are made from stainless steel, with the main valve seat being stellited cast steel. Instead of a gland using some form of asbestos packing to prevent leakage around the spindle, some valves use a labyrinth form of packing with a steam leakoff to the gland condenser or high-pressure drains tank.

Turbine expansion arrangements

In turbines of all types, arrangements must be made for the casing to expand and contract under the influence of fluctuations in steam temperature. In a large number of designs the rotor thrust collar and the associated thrust block are placed at the for'd end of the turbine and this locates the rotor in the turbine casing. The flexible coupling between the rotor and the primary pinion allows the rotor to expand under the influence of temperature change without affecting the axial position of the pinion. The turbine casing is usually supported on a pedestal from the tank tops at the after end or attached to brackets fixed to the gear box. The supporting palms here hold the casing firmly in position relative to the gearbox so that there can be no fore and aft movement between these, male and female spigots with elongated bolt holes being used for this as in type 1, Fig. 21. The spigots and elongated holes, however, do allow thwartships expansion to take place, and care should be taken to see that the spigots are kept clean, greased (molybdenum disulphide grease is usually used) and clear of paint. The spigots may be of dissimilar metals to prevent seizure. At the for'd end

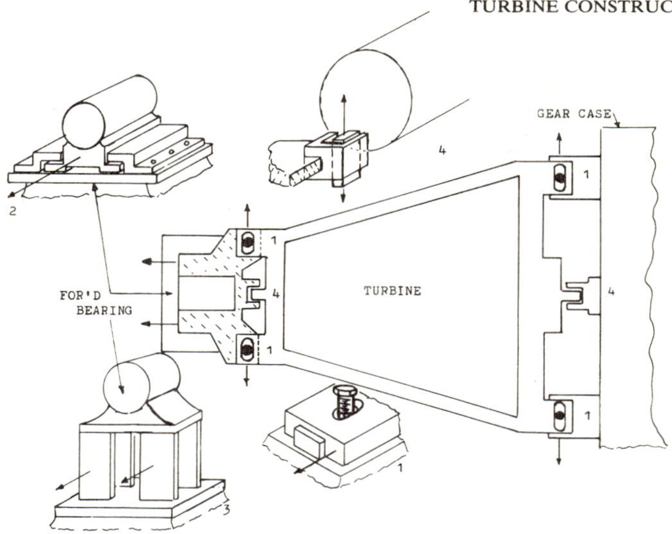

Fig. 21 Turbine expansion arrangements

similar palms are used to attach the casing to the for'd bearing and thrust housing. These palms may use male and female keys or spigots as at the after end with the keys allowing thwartships expansion, whilst expansion in the longitudinal direction is catered for by a sliding foot arrangement or panting plates as shown in Fig. 21 (2) and (3) respectively. As an alternative to the male and female spigot at the for'd end, the palm may simply have an enlarged hole with a stud passing through fitted with a nut and washer giving about 0·254mm clearance between the palm and washer. This then allows expansion in both longitudinal direction and thwartships, but the sliding foot or panting plate arrangement will still be used to provide longitudinal expansion.

To hold the casing in transverse or axial alignment with the gear box, vertical keys as in Fig. 21(4) are fitted at the for'd and after ends of the casing, and half-way along in some cases. These allow vertical expansion to take place, but they must be a good fit to ensure no misali'¸nment takes place.

Pipes fitted to the casing should have large bends or be fitted with bellows pieces and be flexibly supported to allow the casing to move freely and care should be taken to see that no pipe is forced to fit the casing flange as this may act as a restraint to expansion and cause distortion and misalignment. When a turbine is being warmed through great care should be taken to see that the expansions allowed for are actually taking place. Indicators are usually fitted to the expansion arrangement at the for'd end to show any movement taking place. Great care should be taken to see that any sliding foot arrangement fitted is kept clean and well greased as seizure here can cause very serious damage to a turbine.

At the for'd end, the bearing housing and the expansion arrangement are usually supported by a pedestal from the tank tops.

Some designs now only use one panting plate instead of the arrangement in (3). It was found that heat radiating from the turbine casing or nearby pipes raised the temperature of the after plate above that of the for'd one and caused the aft one to expand a greater amount. This then caused misalignment by tilting the bearing.

CHAPTER 3

Turbine Operation

Rotor axial clearances. Procedures for checking when stopped and running.

In the first case a finger plate gauge is used. This consists of a suitably shaped flat steel or brass plate which is firmly attached to the turbine casing between one of the bearings and the adjacent gland. The end of the plate comes close to a suitable collar on the shaft so that the position of the rotor relative to the casing can be checked by taking the clearance between the gauge and collar with feeler gauges. A typical arrangement is shown in Fig. 22(a).

To check the clearances when running a spring loaded spindle is attached to the end cover of the turbine in way of the forward bearing. By lightly pressing the spindle on to the rotor against the spring load the clearance can be checked against a reference mark on the spindle housing which should line up with a similar mark on the spindle. The arrangement is shown in Fig. 22(b).

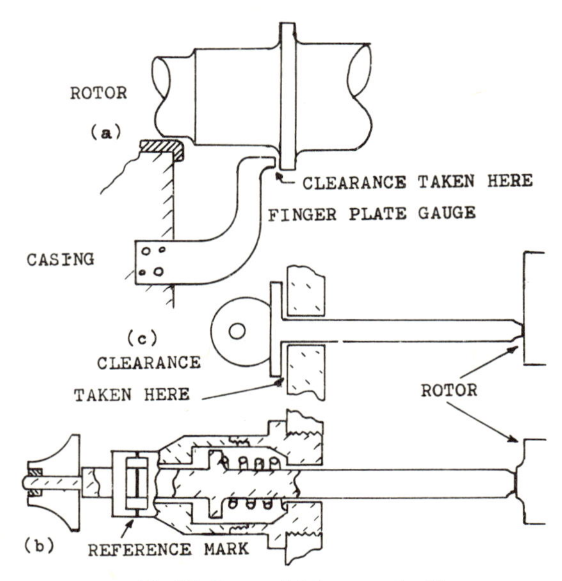

Fig. 22 Rotor axial clearance checking

Another method is to use a poker gauge as shown in Fig. 22(c). This is a steel rod with a machined flange at one end which is passed through a hole in the end cover to the for'd bearing. The gauge is pushed lightly against the rotor and feeler gauges are used to check the clearance between the flange and the end cover. This clearance is then indicative of the rotor blade and/or dummy clearances. When the gauge is removed the hole is filled by a screwed plug. This gauge does not necessarily have to be entered on the centreline of the rotor.

Warming-through and manoeuvring

There are a number of procedures for warming through turbine plants, each depending upon the particular type of machinery involved and circumstances. In general, it may be taken as a rule that the larger the physical size of the turbines, the longer should be the warming through period. The engineer officer in charge of the plant should use his discretion with regard to the time taken for warming through as it will obviously take longer to warm up machinery that has been idle for several days than to warm up a plant that has been operated perhaps 36 hours previously. For the older Parsons reaction type turbines with solid rotors, not less than 4–5 hours should be allowed, unless in a case of emergency, but with the smaller, modern rotors, this time may be considerably reduced. The manufacturer's recommendations should always be strictly adhered to.

The object of warming through is to raise gradually the temperature of the rotors and casings to an even level so that when the turbines are started, irregular expansion or distortion, excessive condensation, which in itself can cause distortion, with associated thermal shocks and excessive stresses are avoided.

There is a tendency for hot steam to flow to the upper parts of the turbine and the condensate to fall to the lower parts when warming through, the upper parts are therefore more likely to become hotter than the lower ones, causing the cylinder to hog. This condition can be aggravated in the LP turbine if a high vacuum and corresponding low temperature is permitted during this period. The rotors also tend to hog in similar fashion if they are not turned frequently. The greatest care should be taken to prevent the localized heating of any part of the turbine and the admission of gland steam should be kept to a very minimum to avoid local overheating of the turbine spindles, just sufficient steam being used to maintain the required vacuum. Similarly, the greatest care should be taken when using astern steam.

There are a number of variations in the methods advocated for warming through a turbine plant, the main differences being in the method of supplying steam for the process to the turbine and the use of vacuum. Probably the best method for steam supply is the provision of small valves which bypass the main manoeuvring valves, just giving an adequate steam flow for the purpose. If these are not fitted, then cracking the manoeuvring valve from its seat or using gland steam as a source of heating steam may have to be adopted. Where the boilers are cold, the boiler stop valves and manoeuvring valves may be opened, gags being used where necessary for this, and steam allowed to pass freely to the turbines as pressure is raised in the boilers. The manoeuvring valves are shut in as the pressure rises to limit the steam flow. A slight vacuum, about 127mm Hg, may be used in all methods to induce a good steam flow without producing a serious cooling effect.

When manoeuvring, the HP turbine casing and manoeuvring valve chest drain valves should be shut for any protracted ahead operation, but reopened immediately this ceases. After 'Full-Away' has been ordered, all drains should be shut and the condensate recirculating valve shut. The turbine should then be slowly worked up to speed, taking at least thirty minutes according to design. As power is being increased always check the turbine casing expansion arrangements.

Bled steam valves may be carefully opened after ensuring the steam lines have been well drained. Great care should be taken to see that the astern manoeuvring valve and double shut off valve are shut tight as any leakage, besides wasting power, can easily overheat the astern turbine. There is usually a temperature restriction of about 120°C for the LP astern turbine, although this figure may vary according to manufacturer, but whatever valve is given, it should be strictly adhered to otherwise the associated distortion may cause a 'rub' to occur.

Warming-through procedures

Prior to warming through the turbines, check that the lubricating oil levels in the drain tank and emergency tank to ensure there is sufficient oil in the system. Top up the gravity tank, if fitted. The presence of any water should also be checked for, and if any is found it should be removed. Also ascertain that there are adequate water supplies to the lubricating oil cooler, although this should not be used until the temperature of the oil leaving the cooler has reached about 30°C.

Open all the turbine casing and main steam line drain valves and make sure that all the steam control valves at the manoeuvring gear and about the turbine are closed, but eased back to prevent the spindles jamming as the turbine temperature rises. All bled steam line drain valves should be opened. Start the lubricating oil pump and see that the oil is flowing freely to each bearing and gear sprayer, bleeding off air if necessary, and check that the gravity tank, if fitted, is overflowing. Turn on the low level alarm for this tank. (On systems with no gravity tank, this step may be omitted and another to satisfy the particular arrangements substituted.)

Ascertain from the bridge that the propeller is clear in preparation for turning the engines. Engage the turning gear and rotate the turbines in each direction to ensure freedom of rotation.

Start the S.W. circulating pump for the main condenser. Start the condensate extraction pump with the air ejector re-circulation valve wide open. Open the manoeuvring valve bypass warming through valves, if fitted. (Cracking open the manoeuvring valve or use of gland steam above may have to be adopted if these are not fitted.) This procedure allows a small flow of steam for heating the turbine and the flow may be assisted by using the air ejector to provide about 127mm Hg of vacuum, without seriously increasing the possibility of distortion. Allow this procedure to continue until the LP turbine inlet belt has reached 71°C (this temperature may vary according to design and the manufacturer's recommendation should be strictly observed). During this time the turbines should be rolled continuously on the turning gear, or if this is not continuously rated, then rolled about every ten minutes, ensuring that the LP rotor is in a different position at the end of each rolling period. Always check the ammeter reading as a high reading gives a warning of trouble, such as distortion causing contact between a fixed and moving component of the plant.

Verify that the expansion arrangements for the turbine are allowing freedom of movement.

Adjust air ejector steam (or condenser air extraction device) and the gland steam so that the minimum gland steam is required to maintain a vacuum of about 254mm Hg in the condenser, taking care not to overheat locally the LP cylinder and rotor spindle with the gland steam. The turbines can be held in this condition until a short period before stand-by. Disengage the turning-gear.

Admit sufficient steam to the ahead turbine to move the propeller shaft about one revolution. Usually a sharp blast of steam is the best way of doing this, closing the valve immediately afterwards, thus avoiding overheating of the moving blades in way of the nozzles on an impulse turbine, which can occur if the steam is admitted too slowly. Do not use astern steam unless it is to check the ahead movement. Repeat this procedure approximately every three minutes for at least fifteen minutes. To check the vacuum raise it to the full operational value, usually 724mm Hg and then drop again to 500mm Hg. The engines are now ready for use.

Whilst waiting for the first telegraph order and during subsequent stand-by periods, the turbines should be turned ahead once every five minutes and if there is any delay the vacuum should be dropped back, gland steam reduced to a minimum and the turbines rolled on steam as previously indicated.

If, during the warming through period, any squeaking should be heard from the turbines as they are being rotated, the steam should be shut off and the turbines left to soak without any movement for at least fifteen minutes.

Certain classes of vessel may adopt variations in warming through procedures according to particular trading requirements, such as rapid warming through for ferries and the maintenance of a high state of readiness, including a full vacuum, for vessels with dangerous cargoes, and the turbines will be designed to suit these conditions. However, strict attention should always be paid to LP casing temperatures and expansion indication, and immediate attention given to any untoward reading or noise. On many modern vessels turning gear operations and steam 'blasting' may be automatically controlled, but the engineer officer should always be aware of the correct procedure in case of an emergency.

Blade erosion

Although the steam supplied to turbines is nearly always superheated, the superheat decreases as the steam passes through the turbine and the last few stages nearly always operate with wet steam. Initially, as the steam passes into the wet stages the water droplets formed are very fine and have the same velocity as the steam in which they are suspended but as the pressure drops further and as a result of further acceleration of the steam, the quantity of water vapour present increases, the particles combine to form larger droplets and they begin to settle out or are thrown out by centrifugal force against the surfaces of the blades or nozzles. Once the solid droplets strike a solid surface, they will never again attain or even approach the velocity of the steam, even though the droplets are removed from the surface by the steam flow. The turbine will be designed for the steam flow speed to match the blade speed so that the steam flows across the blades with minimum friction. If in these LP stages the speed of the water droplets is below that of the steam flow, then they will not pass on to the blade smoothly, but being slower will be struck by the back of the leading edge of the moving blades. This impact will produce a retarding effect on the blades, and some of the water will adhere to the blades, and under the influence of centrifugal force move outwards towards the tips. This action takes place over the whole length of the blade and results in an ever larger concentration of moisture towards the tips of successive rows of blades. Some of this water is thrown from the blade tips and is trapped in narrow annular water traps in the casing, the remainder of the water passes through the blading to exhaust. Although the retarding effect of the water droplets striking the back of the rotor blades is a propulsive loss, a far more serious effect is the erosion or wearing away of the back of the leading edge so that this section of the blade may have a lace-like appearance, with a roughened surface initially, developing into a mass of holes until whole sections disappear. Eventually the blade may fail due to a weakened section.

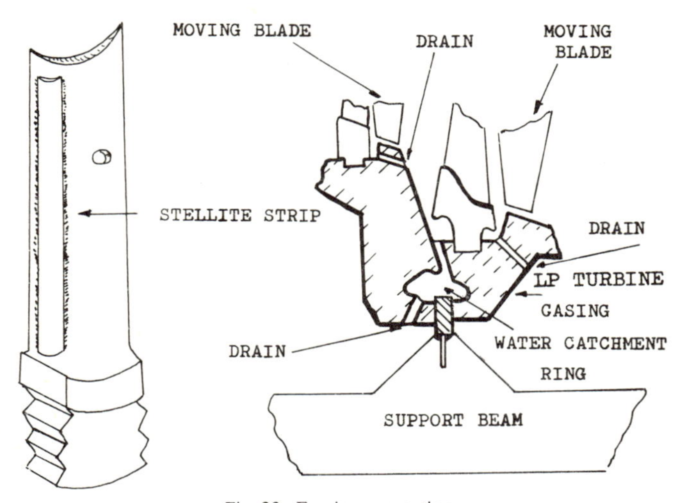

Fig. 23 Erosion prevention

Besides providing water catchment rings in the casing, damage to the blade can be reduced by brazing or electron beam welding on a stellite strip (cobalt, chrominium and tungsten) down the back of the leading edge. Erosion has been known to undermine these strips so that they are thrown off causing serious damage. Other methods adopted involve the use of reheating the steam after the HP turbine exhaust to the initial steam temperature so that the expansion is completed at the dry saturated steam condition.

The LP astern turbine when running in the ahead direction tends to act as a compressor, drawing in any water droplets that may drip from surrounding casing or shields. These droplets can promote erosion, on both leading and trailing edges.

In some cases the blade speed is kept below a certain value, dependent on blade material, so that the impact velocity of the water particle does not create an erosion condition. Operation at low superheat temperatures, high vacuum conditions and the blockage of drainage holes are likely causes of increased erosion. Fig. 23 shows typical arrangements.

Static and dynamic balancing

However much care is taken in the manufacture of a turbine rotor, during the machining and blading processes a slight discrepancy in the tolerances of a disc or blade is bound to occur. Such a discrepancy provides a mass of metal on a particular section of a disc or blade which is heavy compared with the rest, and if the rotor is placed on a pair of level, parallel, knife edged rails, then it will roll along the knife edges until it stops with the heavy section at the bottom centre. To offset this tendency either another mass can be placed on the opposite side to the one causing the trouble to give an equal and opposite correcting moment, or a hole can be drilled on the same side in a disc, the mass of the material removed at a particular radius providing the balancing moment. In Fig. 24(a), if P is the out of balance mass of a particular disc causing the trouble, then the effect can be overcome by placing another mass Q on another disc to provide the balancing moment. The alternative is to remove some material say at point R. This is *static balancing* and for this purpose it

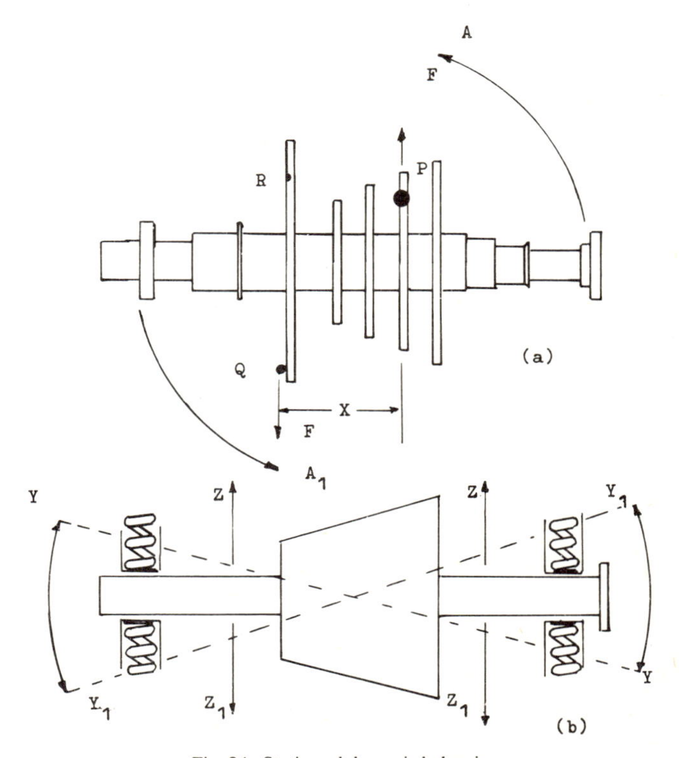

Fig. 24 Static and dynamic balancing

does not matter where along the length of the rotor *Q* or *R* occur. However, if a mass is placed at *Q* to balance the effect of *R* when the rotor runs at speed, the two masses will set up a centrifugal force *F* acting radially outwards at a distance *X* from each other producing a couple tending to rotate the rotor in a direction shown by the arrows A and A_1. Such a couple could cause severe vibration in a turbine running at speed and it is the purpose of *dynamic balancing* to ascertain where the balancing masses should be placed to prevent such vibration occurring. Obviously in this simple case if *Q* or *R* were used on the same disc as *P*, the couple would be destroyed, but the static balance retained. On a large turbine, the out of balance masses together with their axial and angular positions cannot be so easily determined, and dynamic balancing machines are used. One such machine involves the use of soft bearings, that is the bearing keeps are spring loaded in the horizontal plane so that the rotor can move from side to side under the effect of any out of balance masses. The rotor is run at a slow speed, driven through a belt drive and flexible coupling by an electric motor. The rotor remains straight and under the influence of the out of balance masses tends to move horizontally from side to side, as shown in Fig. 24(b), in the direction of arrows Z and Z_1. This is the equivalent of static unbalance of the rotor. It will also show dynamic unbalance by vibrating in a transverse plane as shown by arrows Y and Y_1. In some machines, for dynamic balance, one end is fixed at the bearings, the rotor run up to speed and the out of balance measured; this end is then freed and the opposite end fixed and the rotor again run up to speed for measurements to be taken, enabling the positions of the balanced weights to be calculated. In modern machines electrical pick-ups are used to measure the vibrations; these readings are balanced electrically and a graduated disc shows exactly where metal must be added or removed to provide the correct balance.

A turbine rotor, however stiff it may seem when stationary, will always tend to sag and at high speeds this misalignment will tend to make the rotor bow. Over a particular small range of speeds this bow will tend to become excessive and a very severe vibration effect will occur. This is known as the 'critical speed' and if the manoeuvring valve is opened wider, the rotor will increase its speed and pass through the intense vibration condition to run smoothly once again. There is generally a barred speed range for the turbine operation which is this critical speed, and the turbine should never be operated even for a very short period at this critical speed condition.

Once the rotor has been balanced it should run with the minimum of vibration and in many modern designs vibration monitors are fitted to detect serious disturbances at an early stage of their development. During normal operation vibration may increase slowly over a considerable period of time and may be due to out of balance forces caused by erosion or the build up of carry-over deposits on the blades. Rapid changes in vibration from normal behaviour indicate a serious deterioration in operating conditions, and may be due to blade failure, heavy build up of carry-over deposits, glands rubbing, or if occurring after a refit, alterations in the flexible coupling alignment may be responsible.

Turbine inspection

With most turbines the first procedure is to remove the lagging from the horizontal joint of the turbine and check the expansion indicators to see that the casing has returned to the normal cold position and is not being held in a distorted condition. Any steam pipes connected to the top half should then be disconnected and, if necessary, sections of pipe line removed. When slackening back on the joints, watch for signs of the pipes springing away from the flanges to which they are connected as this indicates that the pipes have been forced to fit the flange and in doing so creating a restraint on the correct expansion of the casing at that point. A local distortion may then be set up which could cause a rub between the casing and rotor. Remove any pressure gauge connections and gland housings, depending upon the type of turbine. Slacken back on the casing bolts, checking at this stage to see if any internal casing bolts are fitted which could prevent the halves separating. This may be done whilst the turbine is still warm, otherwise contraction of the metal may make this difficult. In some cases special bolt heating elements may be used to facilitate this process. As the bolts are eased back, watch for signs of the joints springing apart, this indicates a distortion and the possibility of steam leakage at that point through the joint. Preferably the casing should not be lifted whilst still hot as it will cool down independently of the lower half and possibly distort, making it difficult

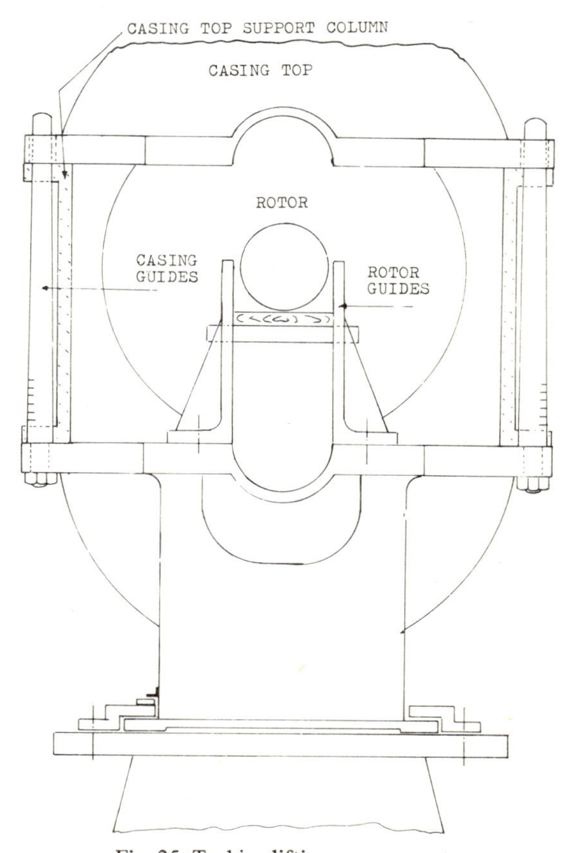

Fig. 25 Turbine lifting arrangements

to obtain a steam tight joint when the two halves are subsequently brought together again.

Guide pillars are now inserted, one at each of the four corners of the casing, passing through holes in the flange of the top half of the casing and being secured to the lower flange by nuts on the underside. These pillars are used to prevent the top casing swinging whilst being lifted, and the graduations on them to ensure an even lift so that the rotor is not damaged as the casing moves up. The lifting gear will depend upon the particular installation and usually consists of chain blocks from beams overhead attached to brackets fitted to the casing into which shackles are fitted. When the casing has been lifted clear of the rotor it may be moved clear of the turbine or supported on columns above the rotor. These columns have flanges at each end and are placed at each corner of the turbine with bolts passing through the flanges securing the columns to the top and bottom casings.

If the rotor is to be lifted, the bearing covers are first removed, then the thrust block covers, together with the pads, then any control gear and finally the flexible coupling is disconnected. The procedure will vary somewhat according to different manufacturers. Before lifting the rotor, the bearing weardown is often checked with a bridge gauge and also the axial movement of the rotor to check for thrust wear.

Guides are now fitted at each end of the rotor in way of the bearing journals, one on each side of the shaft at each journal. The inner edge of these guides is shaped to fit into the journal in way of the fillets so that fore-and-aft movement is prevented, whilst the guides themselves prevent any thwartships movement. Slings are then passed round the shaft, padded to prevent any scoring, and attached to chain blocks hung from overhead beams. The rotor is then slowly lifted clear of the bottom casing, ensuring that it does not foul any of the glands or blading, and then it may be supported on steel bars placed through slots in the guides. Wood is usually placed on top of these bars for the rotor to rest on.

The LP turbine, if of double casing design, has to have two joint faces broken before it is possible to separate the two halves. The outer casing joint external bolts have to be removed and before this can be lifted care should be taken to see that the design does not involve a number of internal bolts. Access to these is usually made through inspection doors in the casing. Once these have been removed the outer casing may be lifted and placed to one side and the inner casing dealt with as previously described. In some cases both inner and outer casings are lifted together and then it may be necessary to fit brackets to support the inner casing from the outer. These are fitted through the access doors and it is essential to check that these brackets are removed when the turbine is boxed up after the inspection as they can seriously impede the flow of steam to the condenser.

The two halves of the HP turbine casing are held together by bolts of creep resisting high tensile steel and in many cases special gear is provided to make sure that all the bolts are stressed to a condition so that the cylinder joint faces remain steamtight when subjected to the full working temperature and pressure. The general procedure is to tighten up the bolts by means of a spanner, slacken them back and then tighten them by hand. A gauge is then placed on the end of bolt and nut and lines scribed across the faces of both. Electric heating elements are then placed in holes drilled in the bolts which are then heated up until the nuts can be turned so that the scribed lines coincide. The elements are then removed, the bolts cool and in doing so compress the joint by the required amount to ensure steam tightness. The manufacturers usually prescribe a particular pattern for tightening up and slackening the bolts.

Fig. 25 shows a typical support arrangement at the for'd end of a turbine.

In some modern designs 'intrascopes' may be inserted into the turbine at selected points to enable periodic inspections to be carried out without having to lift the top half cylinders.

Operating damaged plant

Figs. 26(a) and (b) show respectively the piping, blank flange and water spray arrangements for a plant with HP ahead and LP ahead and astern turbines.

In the event of damage to a rotor or its associated primary reduction gearing to the extent that it is impossible or unwise to continue operating the damaged unit, means are provided to enable the vessel to continue the voyage by steaming on the remaining units. Such an occurrence involves the disconnection of the affected rotor, or if it is a pinion, then the complete removal of this unit. The vessel is usually supplied with special portable pipe lengths, blank and ring flanges, orifice plates and water sprayers so that the system can be modified to suit the conditions prevailing at the time. In general it will be necessary to isolate the damaged unit by fitting the inlet and exhaust pipes with blank flanges, blanking off any 'self-draining' drains, shutting bled steam lines and fitting relief valves where necessary. In this particular arrangement, blanks are placed at points A and D to isolate the HP turbine, all drains are either shut or blanked off, bled steam valves shut off and the emergency steaming pipe sections fitted as shown in Fig. 26(a). These pipes will either be of reduced diameter or have orifice plates fitted to limit the pressures to those allowable in the working units. Depending upon the design, relief valves may have to be fitted

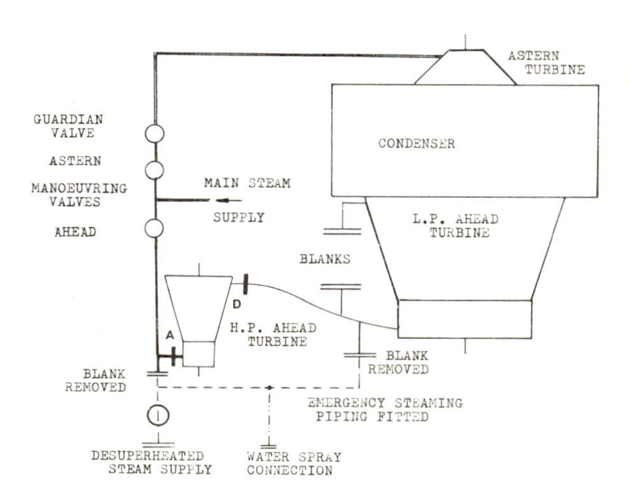

Fig. 26(a) Emergency piping, HP turbine damaged

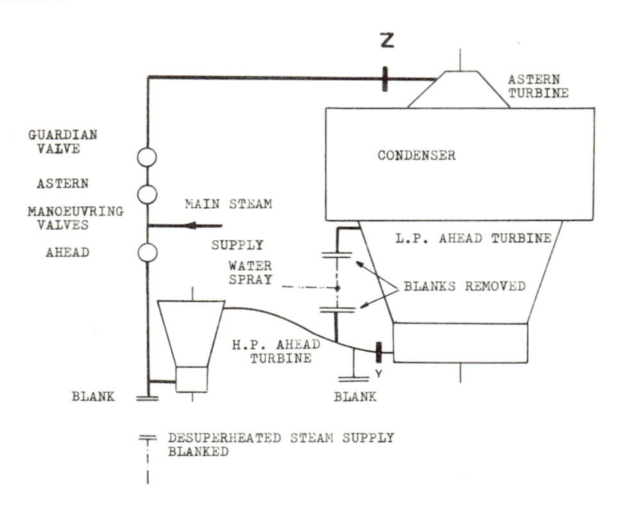

Fig. 26(b) Emergency piping, LP turbine damaged

to safeguard the LP turbine against overpressure and water sprays to keep the steam temperatures down. The water is usually supplied from connections to the extraction pump. Desuperheated steam via a control valve may be used.

Fig. 26(b) shows the flange and piping arrangements if the LP turbine is damaged. Blanks are placed at points Y and Z to isolate the turbine and the emergency piping (as shown) is fitted. This enables the exhaust from the HP turbine to pass directly to the condenser. Fit relief valves on the HP end of the exhaust pipe, if necessary.

A restriction is provided to limit the exhaust pipe area from the HP ahead turbine to ensure that full vacuum cannot act on the last stage diaphragm of this turbine and so overstress it. Full vacuum should be maintained in the condenser when running ahead to prevent the condenser from overheating and water may be spayed into the emergency exhaust pipe where it enters the condenser. Restrictions in the pipelines will protect the condenser, but it may be necessary to fit a relief valve if there is not already one present. Operating procedures will depend upon the degree to which the superheat temperature can be adjusted. All LP bled steam valves should also be shut, and seals fitted to the LP shaft glands.

As the plant will be operating under abnormal conditions with one of the turbines out of action, great care should be taken to watch turbine and condenser temperatures and pressures, as well as those for the rest of the plant. Any signs of abnormal values should be carefully checked and operating conditions adjusted to bring them into line with those normally found.

If only the turbine is damaged, either HP or LP, only the flexible coupling between the affected rotor and its pinion need be disconnected. However, by disconnecting one rotor the torsional vibration behaviour of the turbine, gear and shafting system will have been upset, altering the critical speed for the plant, and the onset of vibration due to this should be watched. Also, the gear train for the affected rotor, will now be idling with only the oil drag of the bearings creating any load. This condition makes the gear teeth susceptible to separation and possible damage. If time permits therefore it is good practice to remove the secondary pinion and so isolate the gear train, fitting blanks to the bearing oil supply lines to this pinion to prevent loss of oil pressure.

Arrangements are made in all turbines to drain condensed steam from pockets in casings, nozzle boxes and pipe lines in which it may collect.

During warming through or manoeuvring procedures steam will condense on cooled surfaces of piping and casings, tending to collect in pockets. This can produce localized undercooling and distortion due to uneven heating which in turn may cause a rub to occur between the rotor and casing. Also slugs of water carried over from

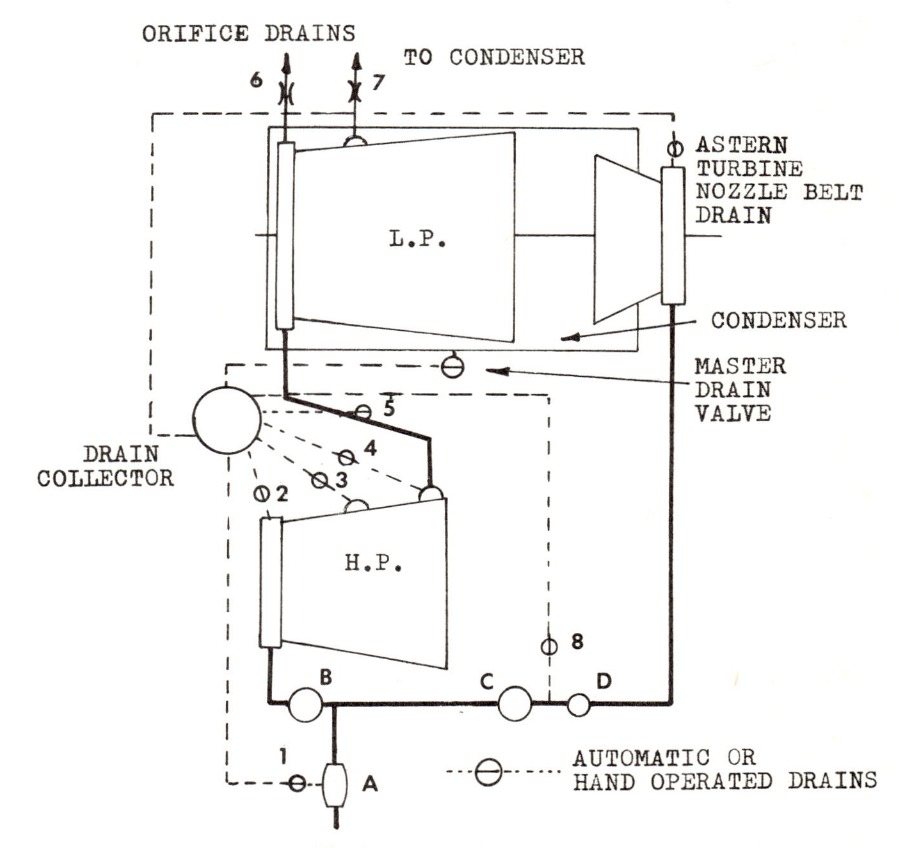

Fig. 27(a) Turbine drain system

these pockets can cause physical damage to the blades as they rotate, and similar damage may occur if the blades strike the water collecting in such a pocket. In many installations automatic drain valves are fitted, so relieving the engine-room staff of the necessity to open and shut valves as warming though and manoeuvring procedures are carried out, but these valves should be checked to see that they are operating properly. It is essential that all drain valves should be opened when warming through and manoeuvring at slow speeds to remove any condensed steam that may form. However, when running for extended periods at higher speeds, considerable quantities of high temperature steam may be bypassed to the condenser from the HP turbine drains, causing the condenser to overheat, and under such conditions these drains should be shut. Staff should of course remember to open them again once the speed drops. At the lower pressure and temperature regions of the turbine installation, such as the LP casing and nozzle boxes, there may be quantities of water forming under normal operating conditions and therefore permanently open orifice plate drains may be fitted to ensure the continuous extraction of any condensation. In some early designs and with some turbo-generators the drains were led to the tank tops, but this tends to cause high engine-room temperatures and humid atmosphere as well as resulting in a considerable loss of water. In most cases therefore the drains are taken to the main reserve feed tank and/or to the condenser. Non-return valves and looped pipes are used to ensure that water can never be drawn back into the turbine casing.

Fig. 27 shows a typical turbine drain arrangement. The drains are as follows:

1. Steam strainer
2. HP ahead turbine nozzle belt
3. HP turbine bleed belt
4. HP turbine exhaust belt
5. HP/LP turbine crossover pipe
6. LP turbine ahead nozzle belt
7. LP turbine bleed belt
8. Astern manoeuvring valve double shut-off drain

A. Steam strainer B. Ahead manoeuvring valve C. Astern manoeuvring valve D. Astern manoeuvring valve double shut-off.

With this arrangement, should an HP drain leak, the master shut-off valve on the condenser can be closed under steady steaming conditions.

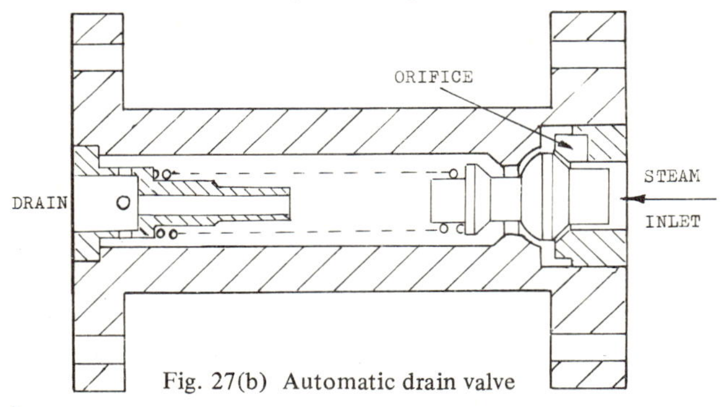

Fig. 27(b) Automatic drain valve

Turbine drain system

The automatic drain valve shown in Fig. 27(b) has an upper and lower valve seating arrangement so that when the steam pressure is low the valve head is pressed by the spring against the upper seat. This has a number of grooves around its circumference to permit free drainage. As the steam pressure increases it acts on the valve head to overcome the spring pressure, push the valve down against the lower seat and thus shut the drain.

Lubricating oil system

The proper functioning of the turbines and gearing is dependent upon a continuous supply of lubricating oil at the correct pressure and of the correct viscosity, temperature and purity. The oil has two important functions to perform in the operation of turbines and gearing, one is to provide a low friction oil film between sliding surfaces and the other is to remove any heat generated in the bearings or

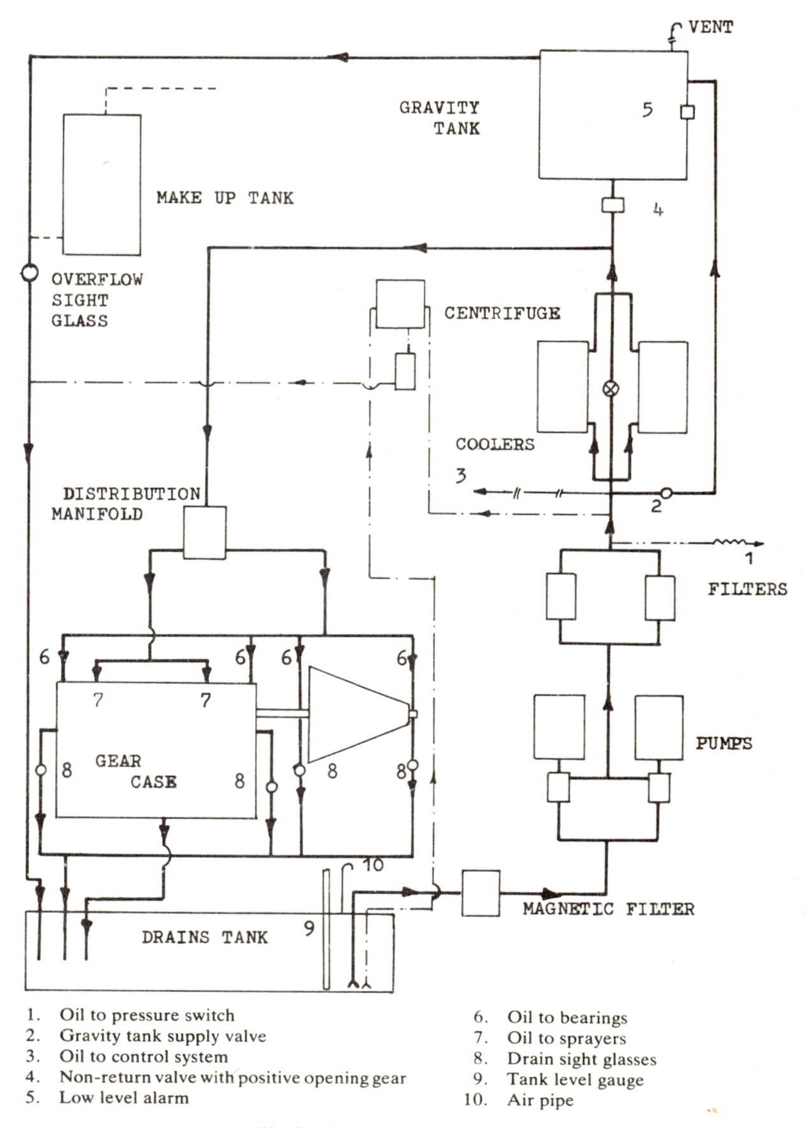

1. Oil to pressure switch
2. Gravity tank supply valve
3. Oil to control system
4. Non-return valve with positive opening gear
5. Low level alarm
6. Oil to bearings
7. Oil to sprayers
8. Drain sight glasses
9. Tank level gauge
10. Air pipe

Fig. 28 Turbine lubricating oil system

conducted along the shafts from the turbine components in contact with high temperature steam.

In this system oil is drawn from the drain tank in the double bottom of the ship by a pump, passing through a magnetic filter as it does so, where any ferrous particles entrained by the oil may be removed before they damage the pump bearings or gears. In some designs the pumps are placed before the magnetic filters, with the actual pump in the bottom of the drain tank. The drain tank is usually fitted with an easily observed float gauge indicating the quantity of oil present, and an air pipe. After being discharged from the pump the oil passes through full flow static filters which may be of the edge type or disposable element design, where particles exceeding approximately 20 microns are removed. Two filters are usually fitted, only one being in use at any one time. In some vessels filtration down to 5 microns is carried out and purifiers may be employed which operate under vacuum conditions to remove virtually all traces of water. From the filters the oil passes to the coolers where the sea water passes through the tubes and the oil over the outer surfaces, the oil being at a higher pressure so that should there be a tube leakage, the oil will not be contaminated. An oil detector may be fitted on the cooling water outlet. The coolers have bypass valves fitted and these should be kept open until the oil has reached approximately 37·8°C, after which the oil temperature is varied by the quantity of sea water passing through the coolers. In very hot climates it may be necessary to use both coolers to keep the oil temperature down. From the coolers the oil passes to a distribution manifold where it divides into two main circuits, one supplying the turbine and gear bearings and the other supplying the gear tooth sprayers, flexible couplings and main thrust bearing. The bearings are fitted with valves to adjust the flow, air escape cocks, and in some cases flow indicators. These valves have holes drilled through the lids so that the oil supply cannot be completely cut off. After passing through the bearings, the oil drains into a small sump and thence into a main leading to the drain tank. Flow sight glasses are sometimes fitted to the outlets from the bearings.

After passing through the coolers, some of the oil is leaked off through a valve which allows a small flow of oil to be continuously bypassed to the gravity tank, from which it overflows via a sight glass to the drains tank. Oil should always be visible in this glass. There is also a supply valve for filling the tank. Prior to the coolers there are take off points for oil to the centrifuge, for the emergency control equipment and for a pressure switch for starting the stand-by pump.

Various means are adopted to ensure the supply of oil to the bearings and gear sprayers in an emergency, i.e. should a pump stop or should there be a power failure. Two independent pumps are usually fitted and if one stops, a low pressure actuated switch starts the stand-by pump. Both pumps are capable of supplying independently the required quantity of oil for the turbines and gears, but should a power failure stop both pumps, the emergency control equipment should shut down the turbines whilst the gravity tank supplies oil for the bearings and gear sprayers for a sufficiently long period for the turbines to come to rest. This tank may contain about 2700 litres of oil and must last for at least six minutes. Pumps may be of the reciprocating type, centrifugal (submerged), screw or gear type, steam or electrically driven or driven from the main turbine plant. Where there is insufficient head room available for a gravity tank, at least two pumps are required and one must be capable of operation if there is a power failure. Some arrangements incorporate an engine driven pump together with an electric pump, and instead of the gravity tank being at the top of the engine room the emergency oil supply is stored in the upper corners of the gearbox structure. Some systems incorporate coalescing filters to remove water.

Shaft glands

Modern turbines use steam packed glands operating upon the labyrinth principle. This consists of projections from the rotor and casing providing a series of narrow spaces through which steam leaks. The steam passing through each annular space between the tips of the fins and the adjacent shaft is subject to a throttling action which reduces pressure and increases velocity. This increase in velocity is dissipated by turbulence and the formation of eddy currents of steam as it flows into the expansion chambers between each segment of the gland. To be effective the incoming jet must not be allowed to flow towards the next opening and pass straight through, hence the stepped design. The labyrinth gland is therefore a series of nozzles, each nozzle progressively reducing the pressure of the steam passing through the gland. By dividing the gland into a number of sections and using reservoir and collection vessels it is possible to have definite known pressure drops over each section. In this way the number of stages can be calculated for a given flow of steam.

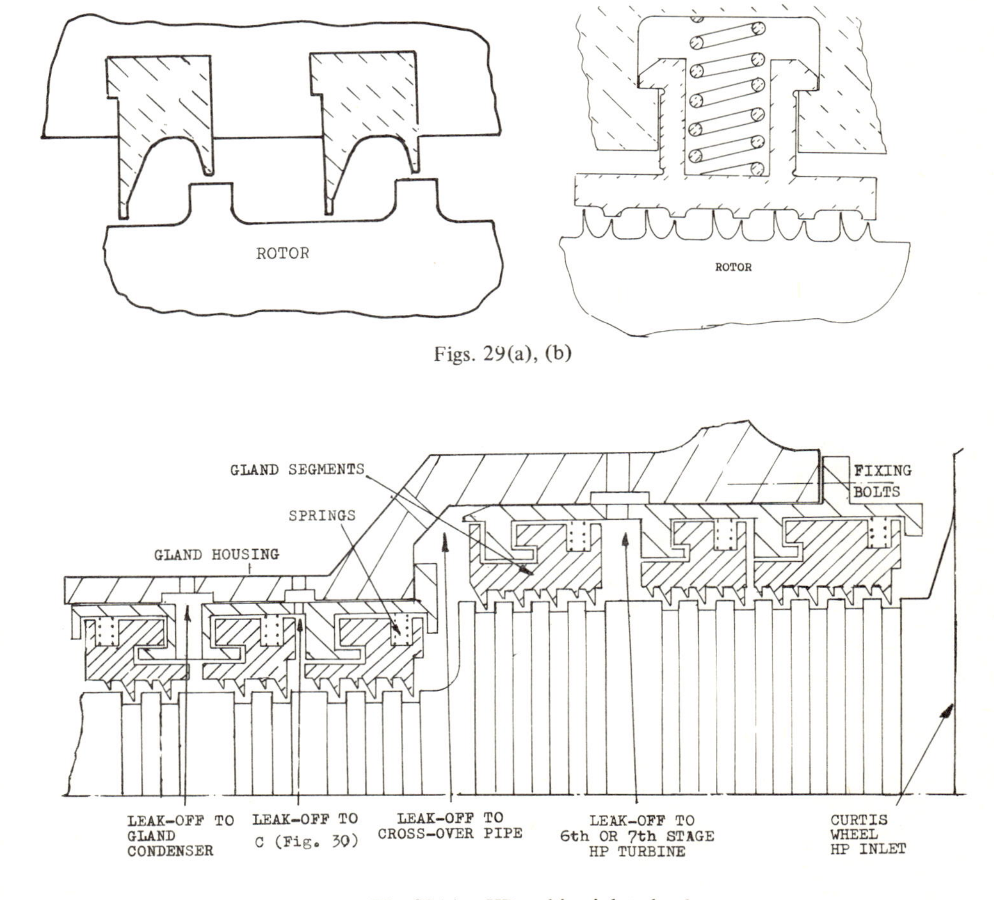

Figs. 29(a), (b)

Fig. 29(c) HP turbine inlet gland

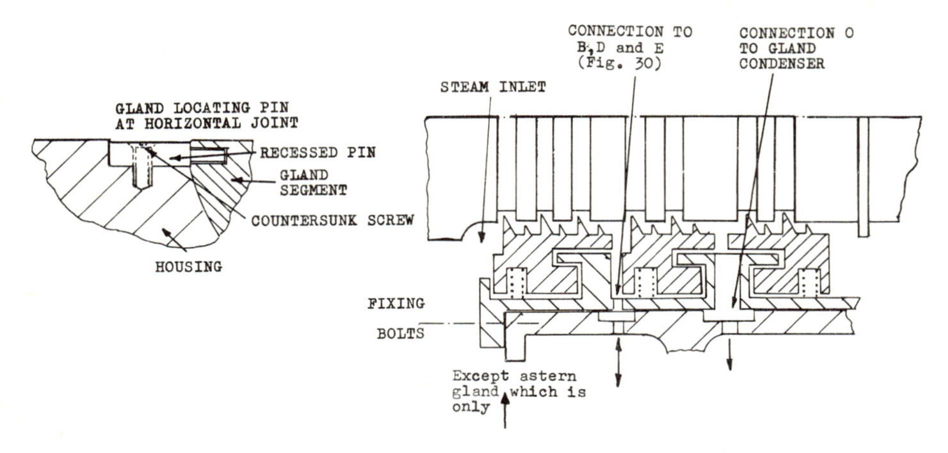

Fig. 29(d) HP exhaust and LP turbine gland

In general the gland consists of a series of fins turned integral with the shaft opposite a series of steps in the casing, or vice versa, although in earlier designs the fins were in the form of strips caulked into grooves turned in the rotor. This design tended to propagate cracks and has been abandoned.

With some HP turbines where there is a considerable pressure drop to be dissipated in the gland and which would require it to be disproportionately long, some of the steam passing through is leaked off to the HP turbine exhaust, enabling the gland to be shortened.

Fig. 29(a) shows a segment arrangement, (c) and (d) show the gland arrangements for the HP inlet and the HP exhaust and LP turbine glands.

Material for the gland segments must be such that if the rotor came into contact with the labyrinth fin, there should be no damage to the rotor. Brass has been used to a considerable extent and copper alloy with a lead content has been found suitable. Lead–copper–nickel alloys can be used with temperatures up to 520°C. Modern turbine designs use a spring backed gland as shown in (b). Radial springs act on each gland segment forcing it into a shoulder in the casing to give the normal radial running clearance of about 0·25 to 0·38mm, between the fin tip and the segment, but if, due to a localized distortion, the rotor and casing come into contact in way of the gland, the spring will give and the rub will not be so heavy as would be the case if the gland was solid, as in (a). With the solid design the fin is thinned away at the tip to reduce the area of contact if a rub occurred. This reduces the amount of heat generated, in the event of this happening, perhaps due to a local distortion caused by carry-over from the boiler or an undrained drain pocket. Contact between the shaft and gland will at first be over a small arc of the rotor surface, and during successive revolutions the heat generated by this will make the rotor bend so that the affected area is pressed further into contact with the casing. This in turn produces more heat and the effect becomes self-accelerating getting progressively hotter until a point is reached at which plastic flow occurs and the material yields in compression. The shaft on cooling then takes up a permanent set. It has been found that the spring loaded gland can deal with such a situation well and has considerably reduced the incidence of shafts bending due to gland rubs.

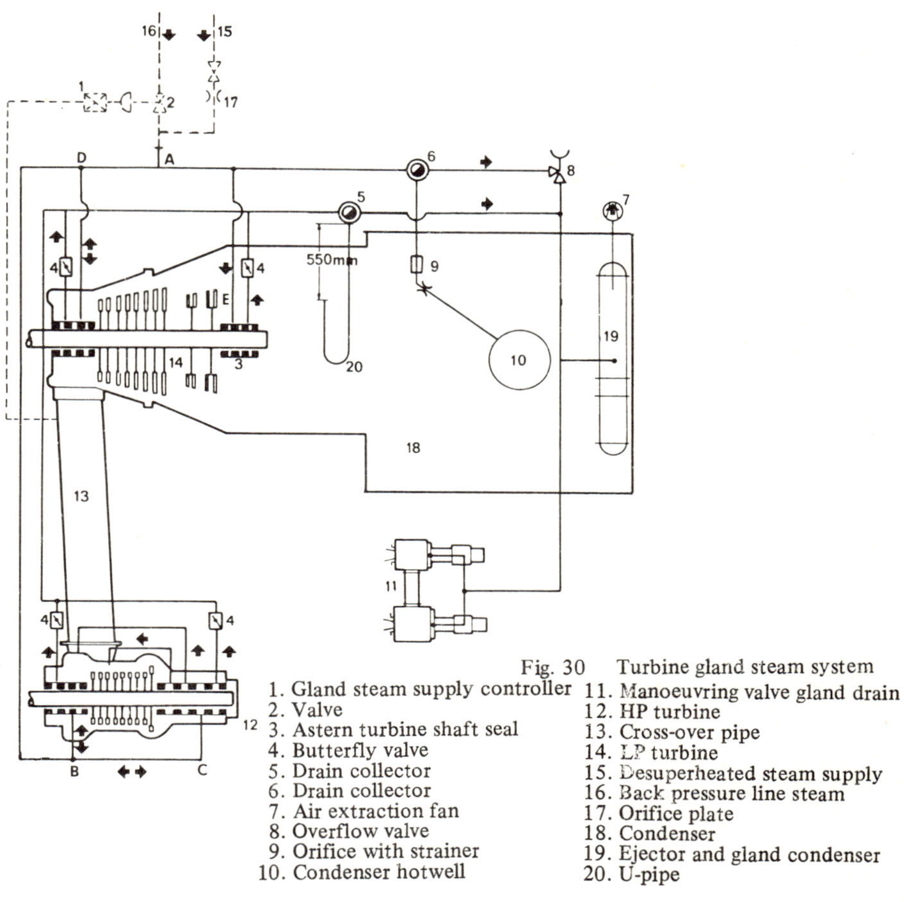

Fig. 30 Turbine gland steam system

1. Gland steam supply controller
2. Valve
3. Astern turbine shaft seal
4. Butterfly valve
5. Drain collector
6. Drain collector
7. Air extraction fan
8. Overflow valve
9. Orifice with strainer
10. Condenser hotwell
11. Manoeuvring valve gland drain
12. HP turbine
13. Cross-over pipe
14. LP turbine
15. Desuperheated steam supply
16. Back pressure line steam
17. Orifice plate
18. Condenser
19. Ejector and gland condenser
20. U-pipe

Figure 30 shows a gland steam system for a two-cylinder, cross-compound turbine in which the HP turbine gland at the steam inlet end has four pockets, as shown in (a), while the remaining HP and LP glands have two pockets as shown in Fig. 29(c). In the HP inlet shaft gland, after passing through the first set of labyrinths steam leaks off to either stage 6 or 7 in the HP turbine to be expanded through the remaining stages. The remaining steam leaking through this gland then passes through the next set of labyrinths where its pressure is dropped and it is leaked off to the cross-over pipe between the two ahead turbines. Here it joins the main steam flow to the LP turbine to be expanded through the remaining stages. The third and fourth leak-offs link up with the main gland steam system for the other glands.

When warming-through or manoeuvring at low power, steam to seal the glands against ingress of air to allow a vacuum to be maintained is supplied from either a desuperheated steam range or a back-pressure line if a turbo-generator set is operating. It is supplied via point A and distributed by way of pipes B, C, D and E.

Under conditions of normal operating powers with high pressure in the HP turbine, the pressure in the LP just above atmospheric and a vacuum existing in the astern turbine on the LP rotor, steam then leaks from the two HP glands and the LP inlet gland through B, C and D to E where it supplies sealing steam to the astern turbine gland, which is under a vacuum and thus prevents air leaking into the condenser.

If the plant is operating below normal power and there is insufficient sealing steam of E, makeup steam is supplied through (2), the quantity being controlled by a valve operated by the pressure prevailing in the cross-over pipe.

When operating at full power there is an excess of steam leaking from B, C and D and this is then bled off through an overflow valve (8). The bleed-off pressure can be adjusted. The bottom of drain collector (6) is connected via an orifice (9) to the condenser hotwell. Steam passing valve (8) is condensed in the gland steam condenser (19) and air is extracted by fan (7).

Excess steam which passes the inner shaft seals is extracted at the outer sealing chambers. The pressure at these points is manually controlled by butterfly valves (4) and should be just below atmospheric. When adjusting the pressure the valves are first closed fully. Steam then leaks through the shaft collars whereupon the valves are opened sufficiently wide to stop this leakage. The

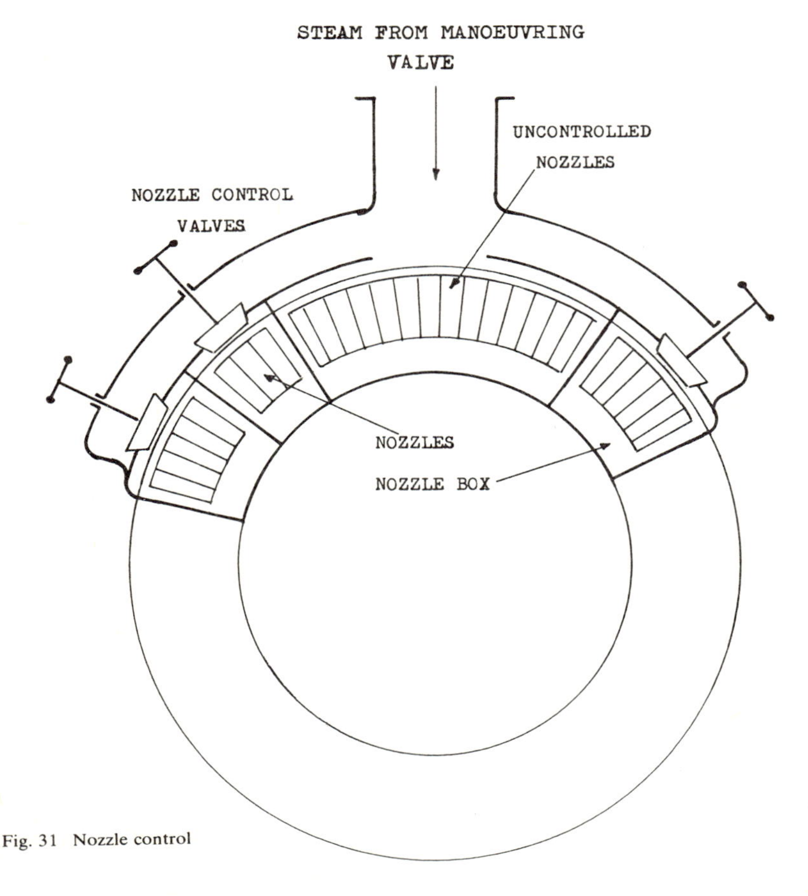

Fig. 31 Nozzle control

extracted steam passes to drain collector (5) which is connected to a U-pipe which should be about 550mm to counterblance the sub-atmospheric pressure caused by fan (7). The approximate gland steam pressure is 1.10 to 1.20 bar. The required sub-atmospheric pressure in the gland steam leak-off circuit is approximately 200mm w.g.

All-round admission, partial admission, overload, nozzle control

All-round admission is a term used describing the flow of steam to a turbine particularly at the HP inlet stages. It merely means that steam flows from the manoeuvring valves to an inlet belt which admits it to a complete ring of blades covering a full 360°. It used to be found on the old Parsons impulse reaction turbines.

Partial admission is a term used in conjunction with impulse turbines. Due to the very low specific volume of the steam at this point, nozzle sizes may be too small for economical manufacture if they were required to cover the whole circumference of the casing. To overcome this problem a smaller number of nozzles of larger size are used covering an arc in either the top or bottom casing, or in some cases in both. The arrangement tends to promote blade vibration as the blades pass through the steam jets issuing from the nozzles. Frequently the section of the casing in which there are no blades has a channel or hood in which the blades run to prevent overheating and loss of power due to windage and the churning action of the blades on steam carried round from the nozzle section.

Ships on certain trades often have to run for lengthy periods at reduced power and to allow this to be done economically the turbines are designed with the nozzles at the inlet to the HP turbines divided into groups, each group having a different number of nozzles. The overall steam supply to all the nozzles is controlled by the ahead manoeuvring valve, but each group, with the exception of one, known as the uncontrolled group, has its own shut-off valve. The uncontrolled group is controlled directly from the manoeuvring valve. The general arrangement is shown in Fig. 31. For manoeuvring, to ensure maximum power availability and to reduce the steam load on the blades, all the individual shut-off valves are opened and the steam to the turbine controlled by the main manoeuvring valve. There have been a number of cases of turbine blades being stripped due to the failure to observe this practice. Under full-away conditions the manoeuvring valve is opened wide and the individual nozzle box valves opened wide in various combinations to provide the desired rate of steam flow for the power required. By employing **nozzle control** in this way the steam supplied at boiler pressure and temperature expands over the pressure drop so that the greatest release of heat possible per unit quantity of steam flowing per unit time is achieved for any power requirements. For vessels on long voyages, nozzle control may be used, but the manoeuvring valve controls a large number of nozzles, and the optimum operating conditions are achieved with this wide open and the boiler and condenser operating at the designed ratings. Should overload conditions be required, extra nozzles are available, controlled by independent shut-off valves.

For **overload** conditions or power conditions in excess of normal a bypass valve is used which admits steam to a belt a number of stages down the turbine from the HP inlet. As the power developed by a turbine depends upon the mass of steam flowing across the blades per second, which in turn is dependent upon the nozzle area and specific volume of the steam, then if the bypass is opened, steam at inlet pressure with a small specific volume is passed to a stage where the blade area for steam flow is greater than that at the inlet to the turbine. A greater quantity of steam passes

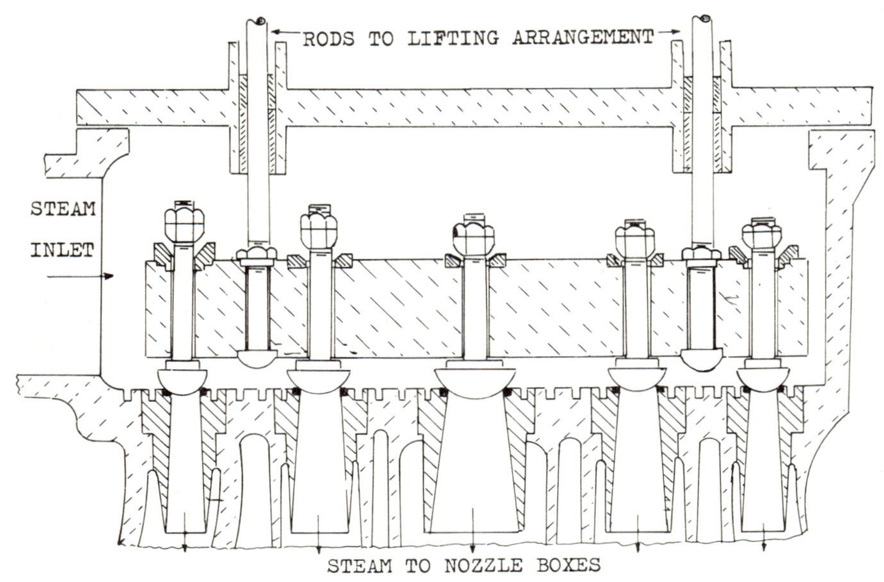

Fig. 32 Sequential control valve

through this stage and successive stages than at normal full power conditions with the bypass shut and therefore the turbine develops more power, but the stages prior to the overload belt may run light and the turbine efficiency is reduced.

Throttling and sequential control

During the course of a voyage there may be passages of considerable length between ports when a vessel is required to operate at reduced power or powers in excess of normal.

The power of a turbine is dependent upon the heat released from a given quantity of steam and the mass of steam flowing per unit time. The heat released is determined by the boiler pressure and temperature and the condenser conditions, and it is common practice to maintain both constant under all variations in output. Power adjustments must therefore be made by keeping the heat released constant and varying the quantity of steam flowing as with nozzle control, or varying both the quantity of steam flowing and the heat released as is the case with throttling.

If the manoeuvring valve is partly shut in, the restriction to flow causes a drop in pressure across the valve, but as no work is done by the steam during this expansion, and as there is no controlled flow as in the expansion through a nozzle, there is no increase in velocity so that the heat content of the steam remains constant, although at a lower pressure, but with a higher degree of superheat. The expansion through the turbine now takes place from a lower pressure and the heat drop available is reduced so that this form of control tends to make less use of the heat drop available and is therefore less economical than nozzle control for reduced powers.

Sequential control is a form of nozzle control used on some modern turbines in place of the hand controlled nozzle boxes of earlier designs. The manoeuvring valve is replaced by two or more valves, each valve controlling a group of nozzles admitting steam to the turbine. As the manoeuvring gear is operated, each valve is opened up in sequence, making a greater number of nozzles available for the steam flow. Fig. 32

shows an example of this type of control valve, and it can be seen that each valve spindle nut has a different gap between the underside of the nut and the collar in the lifting beam. As the beam is lifted, successive valves are opened, exposing extra nozzle groupings to the steam flow and thus increasing this. The expansion across the individual nozzle groups is the maximum possible, i.e. from designed boiler pressure and temperature to the design condenser conditions, making use of the greatest possible heat drop per unit quantity of steam available. In this way the maximum possible efficiency is achieved for all powers.

With other designs the valves are opened by cams fixed to a cam shaft driven by hand or power operated equipment. As the cam shaft rotates successive values and nozzle groupings are opened up to the steam flow. The cams will be characterized, i.e. their profiles adjusted, to give adequate power requirements for manœuvring, particularly at the lower rpm conditions.

The operating circumstances of some modern vessels make the need for economy at low powers unnecessary and such sequentially operated valves are not required. A single ahead manoeuvring valve may then be fitted supplying steam to a group of nozzles, with one or two small groups of extra nozzles for extra power if abnormal circumstances require it. Both ahead and astern manoeuvring valves are designed to give a lift to flow relationship that is linear over the whole range of steam flows. These valves are hydraulically operated using oil at a pressure of about 1034 kilo-newtons per square metre and incorporate a cam operated feedback arrangement which is rapid as the valve starts to open but slows as the valve opening increases. The design of the valves combined with this feedback provides a linear relationship between propeller rpm and actuator movement so that there is sensitive control when manoeuvring at low powers. A simple governor provides speed control for the ahead valve, holding this constant at any desired value over a range which is from about 70 per cent speed upwards. Fig. 33 shows the comparative losses for the three forms of control. Curve (A) represents a single manoeuvring valve with all hand control valves open. Curve (B) represents a single manoeuvring valve with correct operation of hand control valves. Curve (C) represents sequentially operated manoeuvring valves.

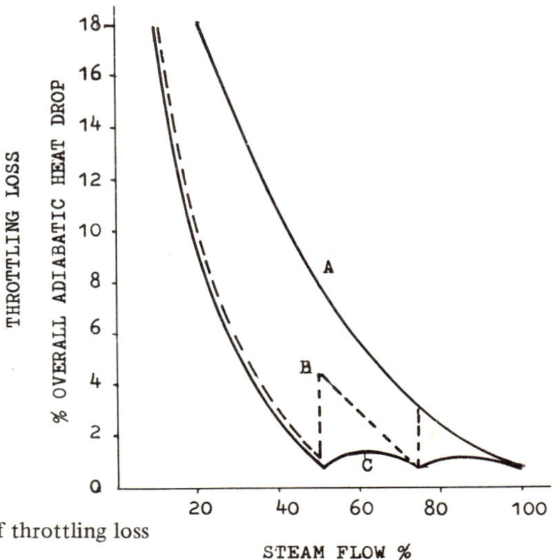

Fig. 33 Variation of throttling loss with steam flow

CHAPTER 4

Turbine Emergency Controls

All turbine plants are fitted with some form of monitoring system which automatically checks the operating conditions and ensures that should any particular component or service move outside its pre-set parameters, then warning is given and the plant subsequently shut down should no corrective action be taken. Fig. 34 shows the basic layout of the protection system and Fig. 35 shows a much simplified circuit.

Tripping devices are provided for components or systems should one or more go into a fault condition:

1. LP turbine axial displacement
2. LP turbine overspeed
3. Second reduction gear wheel end movement
4. Turning gear engaged
5. Low lubricating oil pressure
6. High condensate level
7. Low vacuum
8. HP turbine overspeed
9. HP turbine axial displacement
10. Manual trip from control room
11. High vibration amplitude

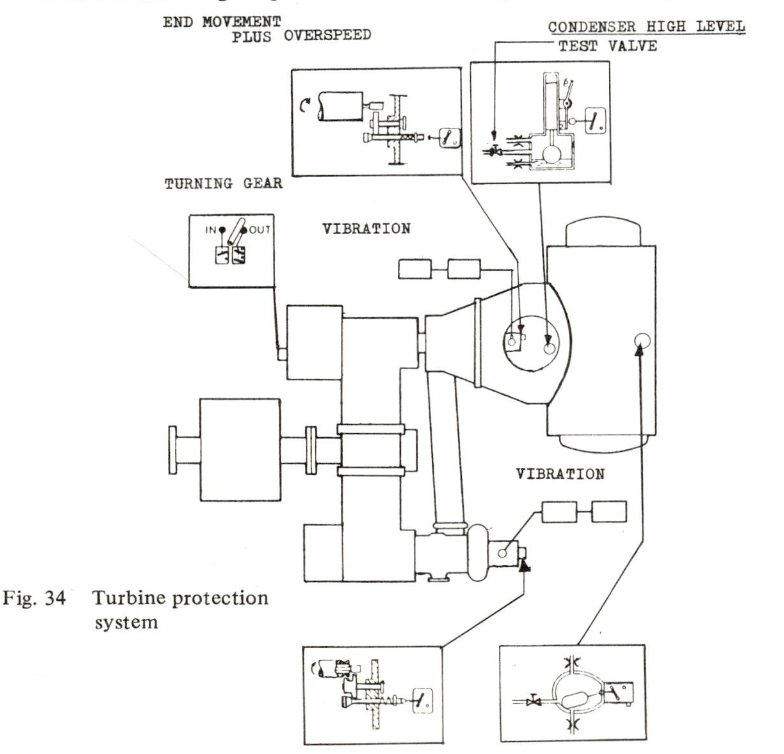

Fig. 34 Turbine protection system

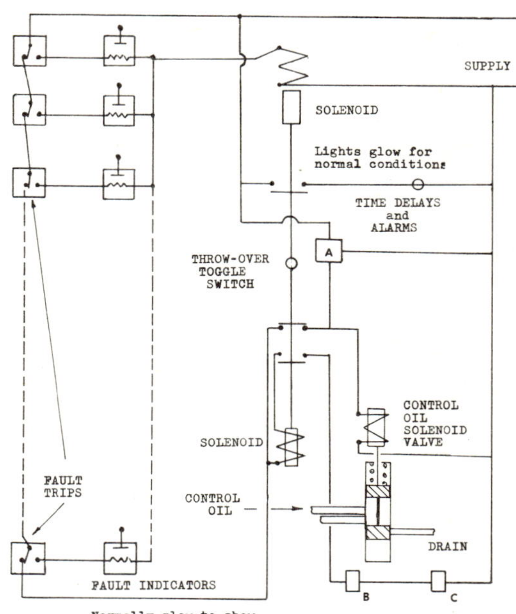

Fig. 35 Turbine emergency trip system

Under normal operating conditions the toggle switch is in the lower position as shown (Fig. 35) with the switch connecting the fault trips to the control oil solenoid valve closed. The fault trips are placed in series so that the supply current must flow through each one before passing through and thus energizing the solenoid valve coil. Should one or more of the protection devices operate due to a disturbance, the trip-switch interrupts the circuit and moves across to close the circuit to the fault indicator and the upper solenoid of the toggle switch. This upper solenoid is then energized and lifts the switch to energize the visual and audible alarm circuits and at the same time break the circuit to the control oil solenoid. With the coil de-energized, the spring forces the spool valve down and allows the control oil to flow to the drain. The drop in pressure in the control oil circuit causes the spring in the manoeuvring valve servo-mechanism to shut in the valve and thus stop the turbines. The upward movement of the toggle switch closes the switch energizing the lower toggle switch solenoid, but the circuit remains broken until the manoeuvring valve positioners are in the closed position. This prevents uncontrolled opening of the valves. After the fault has been rectified, the circuit is reset by push button C, which allows the lower solenoid of the toggle switch to be energized and the relay assumes its normal position with the alarm lamps returning to the glow condition, and with the control oil solenoid valve drain closed.

For manual testing the switch A is operated by a special key which allows the system to be tested without allowing the solenoid valve to operate. Switch B is closed when the manoeuvring valves are shut.

Manoeuvring valve servo motor

Fig. 36 shows the basic layout of a servo-motor for a remotely operated manoeuvring valve. The spool valve is operated by the control system and speed governor via a hydraulic, electric or pneumatic motor which adjusts the position of this valve. If the spool valve is moved up, oil from the control oil circuit enters the space between the two pistons and flows through the ports in the sleeve and the casing onto the top side of the servo-piston which is connected to the manoeuvring valve spindle. The piston is then forced down against the spring, opening the manoeuvring valve. This movement operates the negative feedback linkage connected to the sliding sleeve on the outside of the spool valve, moving this sleeve up until the lower edge of the sleeve top port passes the bottom edge of the spool valve upper piston. The oil flow is now shut off and the servo-piston is held in its new position.

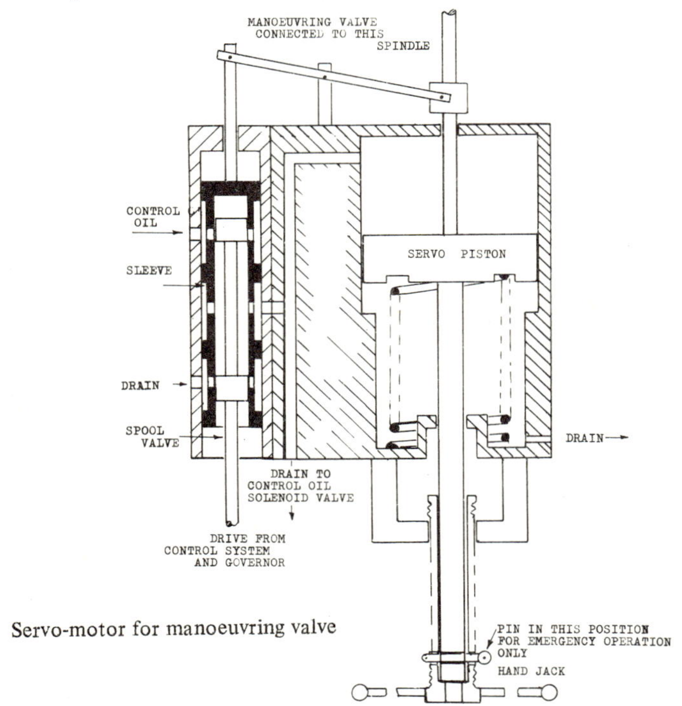

Fig. 36 Servo-motor for manoeuvring valve

If it is required to shut in the manoeuvring valve, the spool valve is moved down opening the lower port, and hence the top of the servo-piston to the oil drain, dropping the pressure above the servo-piston so allowing the spring to move the piston upwards. This movement operates the sleeve linkage so that the sleeve moves downwards until the top edge of the lower port passes the top edge of the lower spool piston. Oil drainage now stops and the manoeuvring valve is held in its new position.

If the solenoid valve in the trip system in Fig. 35 is operated, oil from the top side of the servo-piston is allowed to drain and the spring shuts the manoeuvring valve.

Normally the pin in the manoeuvring valve spindle would not be in the position shown. It is only in this position when the valve has to be operated by hand in an emergency.

Guardian valve for remote operation

Normally the pressure drop across this valve would be minimal as the drain between the astern manoeuvring valve and this valve is connected to the condenser, while the outlet from this valve is connected to the astern turbine which is also under a vacuum. Oil pressure acting through C holds the valve shut. When the astern manoeuvring valve is operated, the oil from the top of the piston is drained away and the steam pressure acting on the underside of the valve forces it open.

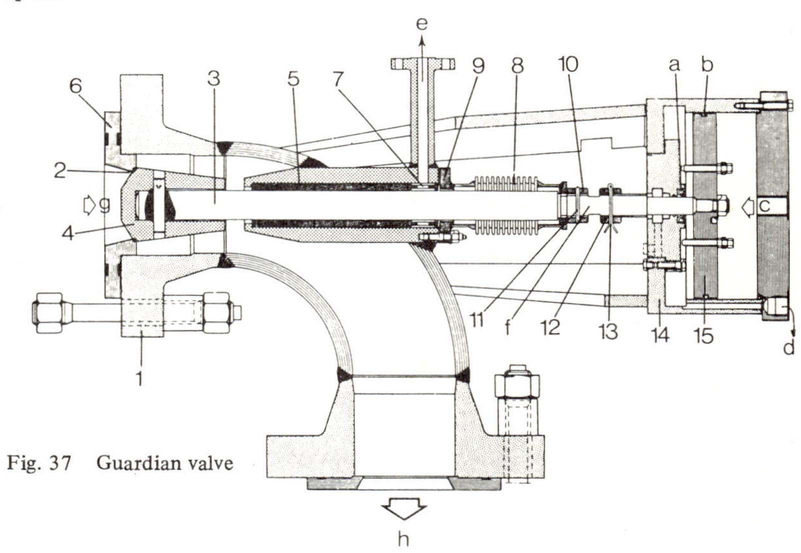

Fig. 37 Guardian valve

1. Valve casing, in the shape of a 90° pipe bend with a welded-in spindle guide. The servo cylinder (14) is welded to the valve casing.
2. Valve seat, clamped between the flange of the Guardian valve casing and the mating flange. The seating surfaces are stellite coated. The valve seat (6) forms the seal between the flanges.
3. Spindle, connected to the servo piston.
4. Valve head, seats on the valve seat (2). The valve head is fixed to the spindle (3) by means of a dowel. The seating surfaces are stellited.
5. Guide bush for valve spindle (3), consisting of five components
6. Valve seat and gaskets
7. Lantern ring for the spindle leakoff steam
8. Bellows assembly
9. Packing, prevents air leakage into condenser
10. Nut to clamp bellows assembly
11. Locking pin for nut (10)
12. Nut for forced closing
13. Split pin for locknut (12)
14. Servo-cylinder
15. Servo-piston

a. Oil sealing ring
b. O-ring
c. Oil supply
d. Drain (1.5mm hole provides a constant leakoff)
e. Spindle leakoff branch
f. Flats for spanner
g. Steam inlet
h. Steam to astern turbine

Forced closing

If the servomotor develops a fault or if the valve spindle seizes, the valve must be forced to close. The split pin (14) is then removed and the spindle is prevented from rotating with a spanner on the flats (f). Turn the nut (13) with a spanner so that the spindle moves in the valve closing direction. The spindle should not be rotated since the bellows (8) would be damaged. When the astern manoeuvring valve is inoperative, the Guardian valve will seal against full steam pressure when closed as described above.

End movement and overspeed monitors

These monitors are placed at the forward bearing pedestals of both the HP and LP turbines. Each monitor consists of two units.

A spring (B) with weight (A) is rolled into the end of sleeve (C) which is inserted into the end of the turbine rotor (D). The spring is deflected by the adjusting screw (E). When the turbine overspeeds the weight (A) is forced outwards actuating the trigger lever (6). This turns anticlockwise allowing the slide (8) under the influence of the spring to move to the left and actuate the limit switch which in turn operates the control oil trip system. The trigger spring (3) must only be replaced by an original spare part as a spring with the incorrect loading can seriously damage the overspeed pendulum.

Also attached to the spindle (2) is the arm (4) with a raised spline which fits into a groove at the end of sleeve (C). Excessive end movement of the rotor causes axial movement of spindle (2) which in turn allows slide (8) to move and thus actuate the limit switch.

The axial position of the rotor may be established by the rod (7) which can be pushed against the end of the rotor by lever (9). The position is then read from the scale and the spring returns the lever and holds the end of the rod from the rotor.

(1) is the end casing of the bearing pedestal.

For testing purposes the unit may be tripped by rod (5) which projects from the end of spindle (2). This is normally covered by a cap to prevent accidental tripping.

Fig. 38(b) shows another end movement monitor in which oil under pressure is forced against a collar from a nozzle. When the rotor is in its correct position the back-pressure of the oil in the pipeline to the nozzle holds a pressure switch in the closed position. Should there be excessive end movement in the rotor thrust, the collar/nozzle gap opens, the back pressure drops and the pressure switch opens operating the trip mechanism for the manoeuvring valves.

Fig. 38(c) shows an electrical end movement monitor which has two coils, one placed either side of a collar on the shaft. Movement of the shaft in excess of ± 3mm alters the inductive impedance of the coils, unbalances the bridge network and trips the alarm and shutdown circuit.

Manoeuvring and cut-out circuit

Fig. 39 shows a manoeuvring and emergency cutout circuit which incorporates a separate pump with a standby pump, bypass valves, filter and non-return valves. The pump supplies oil at 10 bar to servo motors (P), manoeuvring system (M) and the emergency cutout system (E).

For ahead movements the control oil pressure (N) to the ahead servo motor increases when the spring plate in the impulse converter (6251) is kept close to

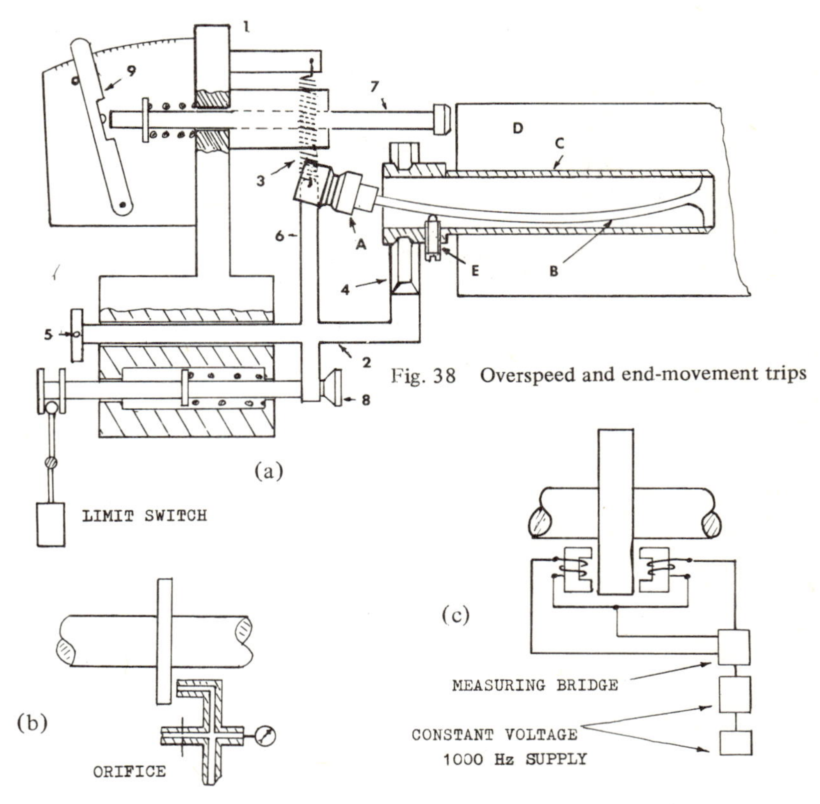

Fig. 38 Overspeed and end-movement trips

the oil nozzle. The pilot piston in the servo motor is depressed and power oil enters the servo. The control oil pressure (N) determines the power oil pressure in the servo cylinder and consequently the speed of the turbine. Power oil is led through relay (6885) to the Guardian valve servo motor.

For astern movements the spring plate is stepped towards the left-hand oil nozzle by means of a stepping motor. The control oil pressure to the astern servo (6312) increases. The ahead valve closed and the Guardian valve relay (6885) changes position and the Guardian valve opens. The astern servo opens when power oil enters the servo cylinder.

An electro-hydraulic speed limiting device limits the propeller speed and is electrically connected to a speed measuring device. If the propeller overspeeds the oil pressure in the speed limiting device decreases and becomes lower than the manoeuvring oil pressure (N). The minimum selector (6881) chooses the lowest oil pressure and the propeller speed then decreases.

In the electro-hydraulic emergency cutout system a tripping signal to the hydraulic solenoid valve (6021) opens the passage to the plungers in the servo-motors. Emergency oil (E) is drained and oil under the servo piston is emptied to the space above the piston simultaneously as the servomotor drains. The manoeuvring valve then closes instantaneously.

During normal operation the plungers in the hydraulic valve (6651) are depressed by the power oil pressure, which through the nozzles and the channels in the plungers acts on the surface above the plungers which are larger than the surface at the other side.

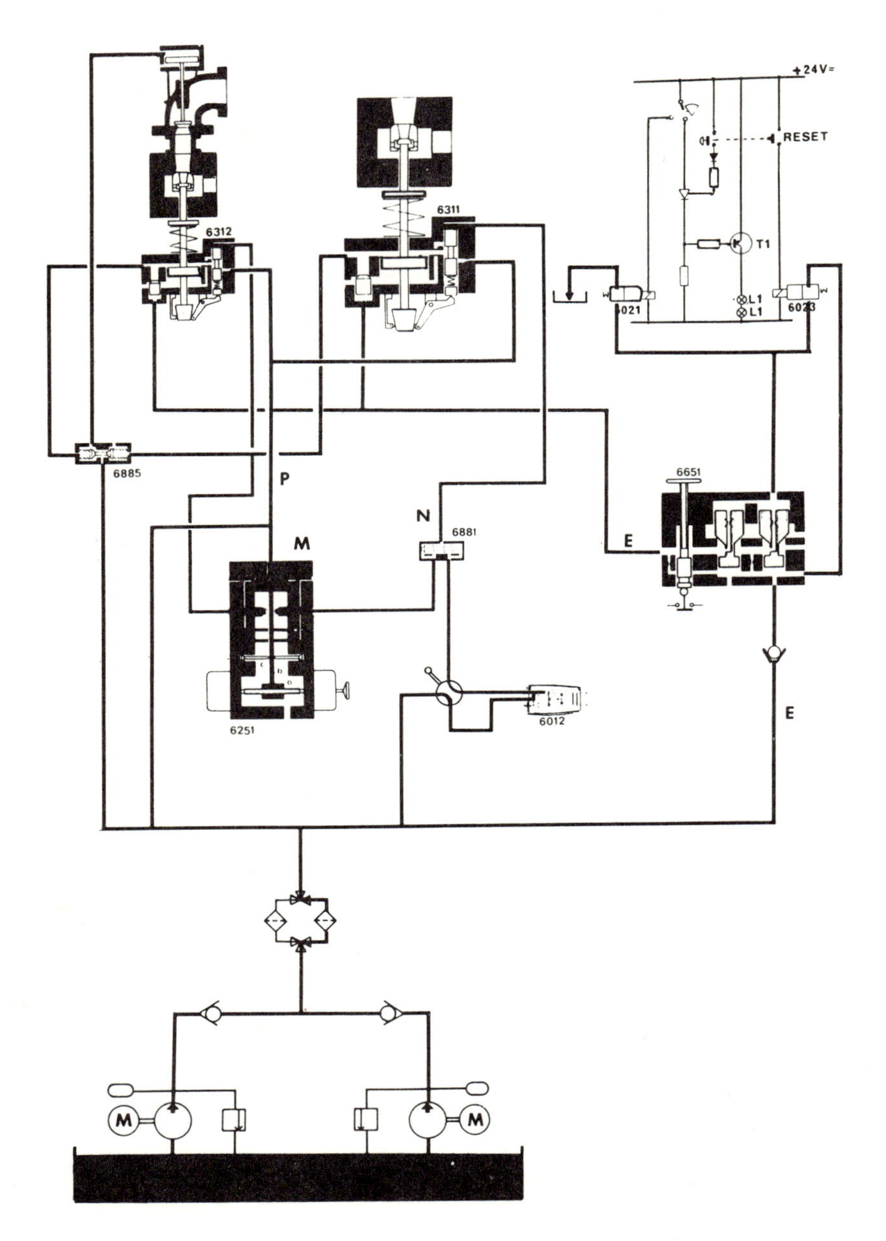

Fig. 39 Manoeuvring and emergency cut-out

CHAPTER 5

Turbine Bearings and Lubricating Oil

Lubricating oil requirements, oxidisation, foaming, white metal bearing corrosion

The bearings and journals of a steam turbine rotor operate under ideal conditions for the provision of fluid film lubrication, speeds being high—in the region of 6000rpm for a modern HP rotor—and bearing loads moderate. For such conditions a low viscosity oil would be suitable, reducing oil film friction and improving the heat transfer behaviour necessary for removing the heat conducted along the shaft from the high temperature section of the turbine.

In the gear case, an oil capable of withstanding heavy sliding loads without shearing the film from the surface of the gear teeth is required, if metal to metal contact and scuffing is to be avoided. Tooth loads on modern gears are high and consequently a high viscosity oil is required to provide the necessary degree of protection.

The oil used in a turbine lubricating oil system is a compromise between these diverse requirements and it has been found that for standard requirements a good quality refined mineral oil derived from a stable paraffinic base stock should give very long service provided care is taken to prevent degradation due to bad operating practices. Where gear tooth loading is exceptionally high, or the quality of the gears poor, an extreme pressure additive may be used to improve the load bearing characteristics. These additives are based on sulphur, chlorine, phosphorous, lead and antimony in various combinations, although these are not in common use at the present time, except for running in purposes. There are drawbacks involved in their use, including lack of compatibility with certain steels and copper.

(a) Under normal operating conditions, a good refined turbine oil should last the lifetime of a vessel, but anti-oxidization additives are usually added to prevent oxidization under the high temperatures that can exist in a modern plant. Air entrainment, moisture and some metals, in particular copper, tend to act as catalysts helping to produce weak acids which tend to degrade the oil so that in time sludges, varnishes and resins form. Once oxidization starts, the oil tends to break down more rapidly and degradation can proceed quickly. Failure of paint on surfaces within the oil system can also promote contamination of the oil. Also, when topping up, not more than 10 per cent of the total working charge should be added at one time as a large amount of fresh oil may precipitate sludge.

(b) Some slight foaming may occur in normal service but prolonged heavy foaming should not arise, and indeed should be eliminated as soon as possible if it occurs as it tends to accelerate oxidization and may affect the operation of control equipment. Adding excessive quantities of fresh oil may be a cause, and air leaks, pump suctions not properly submerged or too near returns and contamination may be contributing factors. Anti-foaming additives may be added to help release entrained air in serious situations.

(c) This is an attack that takes place on tin-based white metal in journal bearings and on thrust pads due to an electro-chemical reaction that produces a very hard dark skin consisting of stannous and stannic oxide on the surface of the white metal. The danger is that the affected surface loses its ability to absorb particles of dirt so that any abrasive substances entering the bearing instead of causing a minor overheating and thus melting the white metal so that they are absorbed, now can score the journal or thrust pad. The hardened surface is also very brittle so that should a piece break away, it can score the journal or pad surfaces. It is thought that water plays an important part, and all possible sources of water contamination should be removed whilst filters capable of removing fine particles of water should be fitted to the system.

Bearing temperatures, emulsification, excessive consumption

High oil and bearing temperatures may be caused by dirty coolers, either on the water or oil side, or both, or insufficient water flow. Lack of lagging on oil pipes near turbine casing, too high viscosity or insufficient oil in the system may be other causes. High bearing temperatures may be caused by misalignment of the rotor or damage to the bearing causing a degree of overloading. Excessive gland steam leakage, the superheated steam raising the temperature of the bearing or washing oil from it can also contribute to high oil temperatures.

Emulsification may be due to water contamination emanating from steam packed glands adjacent to bearings or condensation in the gear case. Contamination with grease, fatty oils, varnish, paint and rust preventatives containing fatty products can also promote emulsification. As the oil becomes degraded through oxidization, emulsions are more likely to form. The presence of an emulsion can be observed by a general cloudiness of a sample and the inability of the oil to free the water in a centrifuge. Salt water emulsifies very easily and is difficult to remove, the most common reason for this form of contamination being the ingress of the water into the system through the coolers when the plant is shut down. Excessive consumption may be due to leakage, and inefficient separation of water, contaminants and sludge in a purifier cause an emulsion which can lead to the loss of good oil as it is carried over to the purifier drain. An incorrect gravity ring in the purifier may also cause a loss. Excessive misting at bearing housings due to high pressures may also be a contributory cause.

Purifiers should be run regularly, preferably for the whole time the plant is running, taking oil from the main sump and returning it to the system. Water and sediment separate from the oil at a rate which varies inversely with the viscosity and therefore the oil is usually heated to about 66°C prior to centrifuging, although the main sump oil temperature is usually adequate. Excessive heating should be avoided and the rate of heating should be carefully controlled to prevent local breakdown of the oil in way of the heater. Watch should be kept on the sludge discharge and if this is excessive, then the oil may need replacing. The most efficient centrifuging rate will depend upon the contamination present and rates will have to be adjusted as a matter of experience. Oil samples should be taken frequently and left to stand to show if any water is present, and if the sample proves to be dark or cloudy, contamination should be suspected. Samples should also be sent regularly to the supplier for detailed analysis.

Turbine rotor thrust block

A thrust block is usually, although not always, fitted at the forward end of the rotor to absorb any axial steam thrust and maintain the rotor in its correct position in the casing. In most cases the pads are of the centre pivot design with either a line or spherical pivot, allowing the pad to tilt as the load is applied encouraging the formation of a wedge of oil between the collar and the white metal on the pad face. It also allows the thrust to accommodate any reversal of thrust that may occur. When running ahead the thrust on the rotor is usually in the direction of the steam flow and load is taken from a single forged thrust collar on the shaft onto a full face of some twelve Michell thrust pads situated in the thrust housing.

A retaining ring incorporating a dovetail shaped groove into which the pads are fitted, is fitted into the housing, the ring being in halves with stop plates at the joints holding the pads in position. To prevent the ring turning with the shaft one of the stop plates is extended into a recess in the casing. Shims and liners are fitted to this carrier ring to enable oil clearance adjustments to be made.

To locate the rotor when running astern a full ring of pads may be used similar to the ahead design, but frequently only a half ring is fitted, this being identical to the

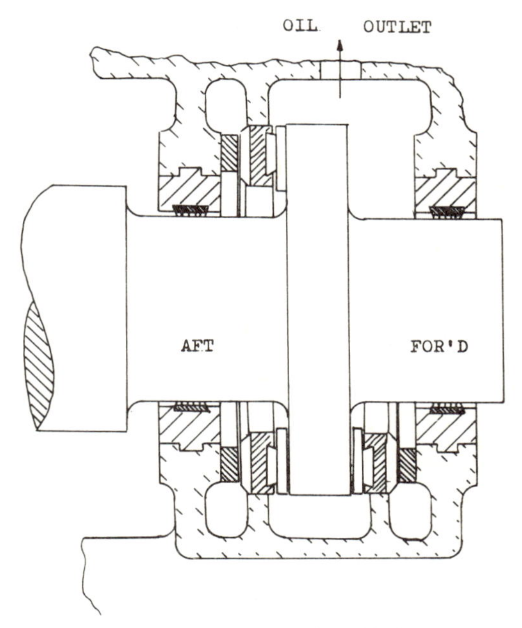

Fig. 40 Turbine rotor thrust block

halves of the full ring. A stop plate is fitted to the casing to prevent the free end of the half ring from lifting.

For lubrication, oil is supplied at the bottom of the casing at each side of the thrust collar with the oil flowing over the pads to the outlet at the top. There is a tendency for the collar to pick up the oil and under the influence of centrifugal force to throw it towards the periphery, starving the inner surfaces and to prevent this an orifice plate is fitted at the outlet at the top. This plate also meters the oil flow from the main oil supply system. Oil sealing rings, white-metalled in the bore to prevent shaft damage in case of contact are fitted to prevent oil escaping along the shafts. Total oil clearances are in the region of 0.254mm in the thrust block and this may be checked by measuring the distance between a collar on the shaft and a finger piece attached to the casing, jacking the rotor forward and aft, when stationary or by pocker gauges when the turbine is running. Gunmetal or mild steel may be used for the backing material on the pads with the babbit metal consisting of 87 per cent tin, 8–9 per cent antimony, 3–4 per cent copper.

In some designs the thrust collar face is case-hardened and fitted to the shaft by a combination of a longitudinal key, an interference fit and a circumferential retaining ring. Fig. 40 shows a typical rotor thrust block.

Turbine rotor bearing

Fig. 41(a) and (b) shows a typical turbine bearing which in modern practice usually has a length/diameter ratio of about two-thirds. The journal/white metal clearance is usually about 0·23mm for a 140mm diameter shaft and shaft clearances are not usually taken below this value. Lubricating oil is supplied under pressure to an annular ring between the bearing keep and shell, passing into the bearing surface through ports cut in the butts. At these entry ports the white metal is chamfered away to spread the oil along the length of the bearing, great care being taken not to cause a ridge at the edge of the chamfer that would scrape oil from the journal and thus hinder lubrication. It is essential that no oilways are provided, lubrication being effected by the rotation of the shaft producing a wedge of oil between the white metal and the journal. The clearance of the bearing is sufficiently large to allow a considerable quantity of oil to flow through to remove heat soaking along the shaft, with the oil draining over the ends of the bearing into a drain well and thence into the drain tank. Great care has to be taken with bearing clearances as if the clearance is too large excessive maladjustment of the rotor can arise, whilst if it is too small, there will be an excessive temperature rise due to the reduced quantity of oil flowing through leading to white metal failure. Bedding in is not required with these bearings as it can prevent oil film build up. A temperature rise of about 55°C can be expected in the oil passing through, with the oil leaving the bearing in the range 54°C to 66°C, but not exceeding 83°C. The white metal thickness varies from 0·25 to 0·5mm, the latter being the better for dirt absorption and taking up slight misalignment. A typical babbit metal for a bearing is 85·5 per cent tin, 8·5 per cent antimony and 6 per cent copper, with the backing of steel or gunmetal. Thicker white metal is used with gunmetal. The end part of each shell is bored out to provide a strip about 0·254mm below the level of the white metal. This strip is intended to support the shaft should the white metal run and thus prevent the gland fins and blades touching the casing. There is a danger of copper pick-up with gunmetal backed bearings should the white metal run, and steel backed bearings, besides eliminating this trouble, seem less prone to de-tinning and provide greater rigidity. Recessed cheese headed set screws at the horizontal joint prevent the bearing from turning in the keep.

An anti-siphon device may be fitted to prevent all the oil leaving the bearing if there is a complete failure of the oil supply and the gravity tank empties.

Bearing clearances may be taken with a bridge gauge or leads and if a bearing should be removed for inspection, care must be taken when lifting the rotor clear to prevent damage to the steam labyrinth packing. A dummy bearing is usually placed in position to support the rotor when a bearing is removed.

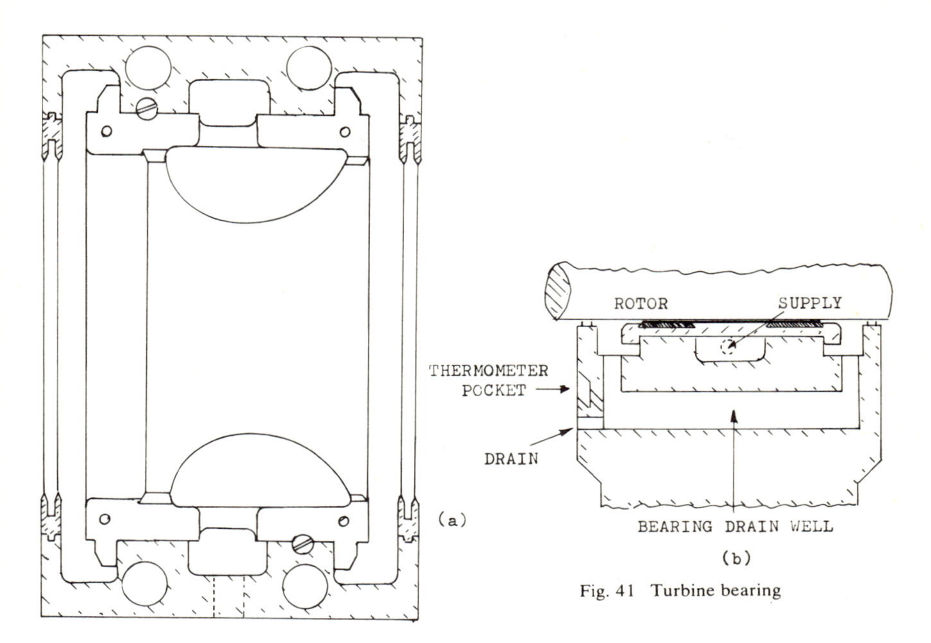

Fig. 41 Turbine bearing

Oil performance

The purpose of the oxidation inhibitor is to retard the chemical breakdown of the oil and a measure of this is the acidity of the oil, usually called the Total Acid Number (TAN). This is the weight of potassium hydroxide in mg which must be added to 1 gram of oil to plate out is acidity. For a new oil the TAN should be 0.05 to 0.15mg KOH/g oil. The figure will depend upon additives and refining procedures, and should be around 0.1 to 0.2mg KOH/g oil for an oil in use. When finally the life of the oil is approaching its end the TAN increases quickly, and when the figure of 0.3 is reached tight control should be kept on the oil as from this level it can reach the limit of 0.5mg KOH/g oil very quickly.

Besides producing babbit corrosion as mentioned previously, water in the oil can cause the failure of the oil film at some highly loaded point leaving no lubrication present, as well as increasing the risk of corrosion particularly to parts above the oil surface. It also contributes to a reduction in the life of the oil particularly in the presence of copper and iron, and if salt is present in the oil there is increased risk of fatigue failure to teeth on gear wheels. If the water content rises to 0.1 to 0.2 per cent an abnormal situation exists and should be investigated. There are a number of tests that can be carried out, the simplest being to take a sample in a glass bottle: if it has a milky appearance this could be due to water or air. If this cloudiness persists after one hour then it is most likely due to water. The quantity that can be thus identified is about 0.01 per cent (100ppm) upwards. Another test is to mix a given quantity of oil with a set quantity of kerosene in a flask and then tightly screw on the cap which contains a water-reacting chemical such as calcium chloride. The gas pressure produced is a measure of the quantity of water present. Test sticks which react with either fresh or salt water are also available.

Contamination with diesel lubricating oil, greases and temporary anti-rust compounds can have disastrous effects on turbine oils, often producing slimy, grease-like compounds.

CHAPTER 6

Turbo-generator Controls

Steam supply control valve. Protection system

Fig. 42 shows an arrangement for a back pressure turbo-generator by which the flow of steam can be controlled to provide the optimum efficiency in operation over a wide range of loads. It also incorporates hydraulically operated safety devices to shut the machine down if abnormal operating conditions occur. In some designs the valves controlling the steam flow to the nozzles are hydraulically operated, eliminating the mechanical linkage shown in this diagram. Whatever the operating arrangement for these valves, the design should allow for the maximum heat drop to be realized by expanding the steam from the pressure at the nozzle inlet belt to the exhaust condition with the minimum throttling for wide fluctuations in load. This is usually achieved by having the operating arrangements for the valves opening them up in sequence as the load increases and thus exposing a greater number of nozzles for the steam to flow through.

The movement of the valves is brought about by a centrifugal governor driven from the rotor operating through an oil operated servo-mechanism. The piston of this mechanism operates a mechanical linkage which in turn moves rocker arms attached to the control valve spindles. These double beat valves adjust the steam flow to the nozzle groupings as shown. A number of devices are provided for the protection of the turbo-generator against abnormal operating conditions. Oil for lubrication and control is supplied when the machine is running by a gear pump driven off the rotor shaft and discharging through a fine grade filter giving filtration down to about ten microns. An auxiliary oil supply from an independently driven pump is usually made available for stopping and starting, together with a hand pump for emergency use. The relay oil passes through a fine filter and then to a combined low oil pressure trip and high exhaust pressure trip (this would be replaced by a low vacuum trip on a condensing turbo-generator). Also on some designs these valves may be independent. The low oil pressure trip protects the turbine against journal bearing, thrust bearing and gear failure due to the loss of lubricating oil pressure. The trip usually consists of a spring loaded hydraulic spool valve which balances the lubricating oil pressure, of which the trip oil pressure is a function, against a spring load and thus allows the relay oil to pass to the steam stop valve and throttle valve relay cylinder. When the lubricating oil pressure falls, the trip oil pressure also drops, the spring overcomes the load exerted by the trip oil on the spool valve which then moves to a position shutting off the relay oil and causing the steam flow to the turbine to be interrupted.

The overspeed trip comes into operation at about 10 to 15 per cent overspeed releasing the trip oil pressure to the drain, which in turn actuates the low oil pressure trip, shutting down the turbine. The trip usually incorporates an eccentric component which is designed to come into operation when the centrifugal force set up by the

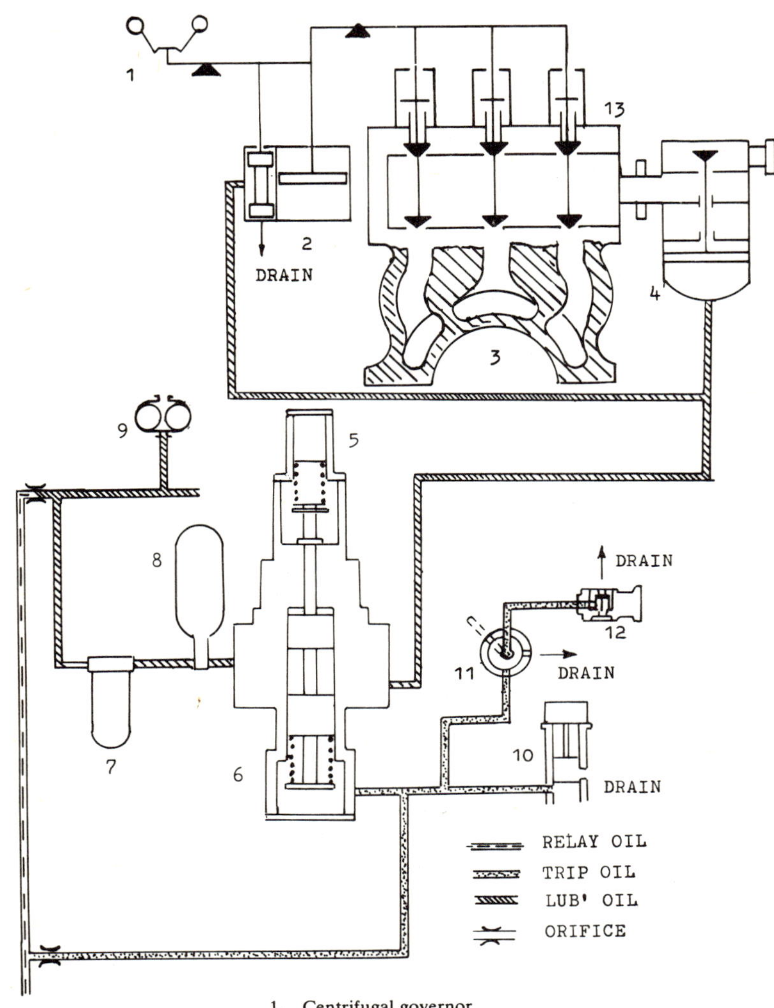

1. Centrifugal governor
2. Oil operated servo-mechanism
3. Casing nozzle boxes
4. Steam stop valve oil cylinder
5. High back pressure/low vacuum trip
6. Low oil pressure trip
7. Relay oil filter
8. Relay oil accumulator
9. Main oil pump
10. Solenoid trip (energize to trip)
11. Hand trip
12. Overspeed trip
13. Steam control valves

Fig. 42 Turbo-generator control system

eccentricity overcomes a spring load and moves a component releasing the oil pressure. A local hand trip releases the trip oil with a similar effect, as does the solenoid trip which is used for shut down from a remote position. A high exhaust pressure trip (or low vacuum) designed to protect the turbine exhaust casing and steam line or condenser shell from excessive pressure and the LP blading from high temperatures is sometimes fitted. This again is a hydraulically operated spool valve incorporating a spring acting on the valve at one end and a bellows fitted to the other. A casing surrounds the bellows and the space between the casing and bellows is connected to the exhaust steam line or condenser. Normally the spring load combined with the bellows loading maintains the spool valve in a position where the relay oil is allowed to pass through to the throttle valve servo-mechanism and the stop valve oil cylinder. If the level of the exhaust pressure should rise or the vacuum drop to a preset level, the change in loading on the bellows causes a state of inbalance with the spring, allowing the spool valve to move to interrupt the relay oil flow to the steam stop valve and control valve servo-mechanism, thus shutting down the turbine. The low oil pressure protection device usually incorporates a locking arrangement that requires resetting by hand. This is to ensure that it cannot return to its normal position automatically once the oil pressure has been restored and restart the turbine without the knowledge of the operating staff.

Turbo-generator governor

Fig. 43 shows a typical centrifugal governor arrangement operating a hydraulic servo-mechanism which in turn adjusts the position of the control valves admitting steam to the turbine. According to the design there may be more than one such valve. The top diagram shows the behaviour of the governor with a decrease in load and hence an increase in speed. Under the influence of the resulting increase in centrifugal force, the weights (W) move out, lift shaft (S) and compress spring (C). At the same time, due to this upward movement, linkage (M) pivots about point (A), moving point (O) and linkage (N) downwards about point (Z). This movement also moves the spool valve (P) down, admitting relay oil to the underside of piston (K), whilst allowing that from the top side of the piston to drain. Piston (K) therefore rises, its spindle also lifting points (Z) and (B) and thus pivoting linkage (Y) about (D). This moves the steam control valve down, reducing the steam flow to the turbine. The upward movement of the piston (K) also tends to lift the spool valve (P), restricting the flow of oil to the underside of piston (K). As the speed of the turbine drops to the required value, the weights (W) move towards the centre, spring (C) expands and the shaft (S) moves down. Linkage (M) pivoting about point (A) moves up, lifting point (O) and linkage (N) and bringing up the spool valve to its mid position where it shuts off the relay oil to the underside of (K) and the drainage from the top side. Piston (K) is held in this new position, as is the steam valve and the turbine operates at its designed rpm under the new load condition.

For an increase in load and thus a drop in speed, the weights (W) move towards the centre as the centrifugal force acting upon them drops, the shaft (S) moves down, spring (C) expands, linkage (M) moves up, pivoting about (A). Point (O) and linkage (N) also move up, pivoting about (A). Point (O) and linkage (N) also move up pivoting about (Z) and lifting the spool valve (P). Relay oil flows to the top side of piston (K), at the same time draining from the underside, so that piston (K) moves down. Points (B) and (Z) move down with it, pivoting about linkage (Y) about (D) and lifting the valve (R) to admit more steam to the turbine. As the rotor speed

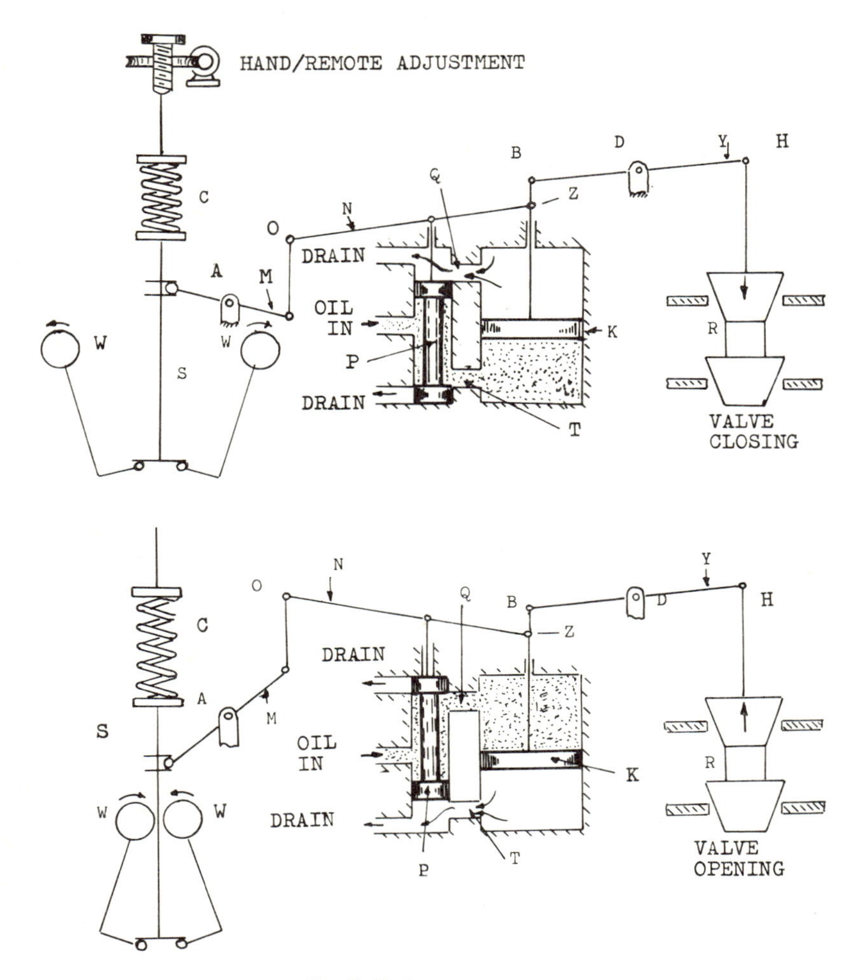

Fig. 43 Turbo-generator governor

increases to its normal value, weights (W) under the effect of the centrifugal force move out, shaft (S) moves up, spring (C) compresses, linkage (M) pivoting about (A) moves down, taking with it point (O) and linkage (N). This moves the spool valve down to mid position, shutting off the relay oil to the topside of position K and the drain from the underside. Piston (K) is again held in a new position, together with the steam valve, so that the turbine operates at its designed rpm once more under the new load condition.

CHAPTER 7

Gearing

Advantages of double helical gearing

In a straight tooth or spur gear, meshing is accompanied by impact as the load is transmitted from tooth to tooth, the average number of teeth in contact usually not being less than 1·2 to 1·4. This is not too troublesome at low speeds, but at high speeds it is not possible to mesh straight teeth without noise or shock. To overcome this problem the teeth are given a twist along their axis to form a helix, hence the name helical gear. The angle of the helix is selected so that one end of the tooth enters mesh before the opposite ends of preceding teeth have disengaged, the angle adopted usually being 30° to the axis of the shaft. In this manner several teeth are in mesh at the same time and a very smooth transfer of power results. This helical form of the gear produces an axial loading which has to be absorbed; this is usually achieved by using a double helical gear arrangement as shown in Fig. 44(a), so that axial loads balance out, as represented by A and A_1. Due to irregularities in manufacture and due to subsequent wear, there may be unbalanced axial forces existing and the gear tends to shuttle. It is important that the flexible coupling is free to allow this to happen otherwise heavy overloading of the teeth may result.

Some early methods of gear cutting resulted in a considerable lack of uniformity of the gear teeth at the point where the hob started and finished at the ends of each helix. To reduce the shock of entry of the teeth into mesh caused by this, each tooth is dressed for a short distance at each end, being cut back at an angle of 45° to 60° to the pitch line and tapered on each flank for a short distance back, as in diagram (b). Tip relief is also sometimes given by dressing back the tips as shown in diagram (c) to reduce the loading and prevent the subsequent breakdown of the oil film. Care must be taken not to give too much tip relief otherwise the effective depth of the tooth is limited, reduction in contact line occurs and the number of teeth in contact at any one time is reduced.

Involute tooth form. Tooth design definitions

The most common form of tooth used in marine reduction gears is of the involute type. In this form, the side or flank of the tooth is derived from a curve belonging to the involute family of curves, which is a curve formed by the end of a

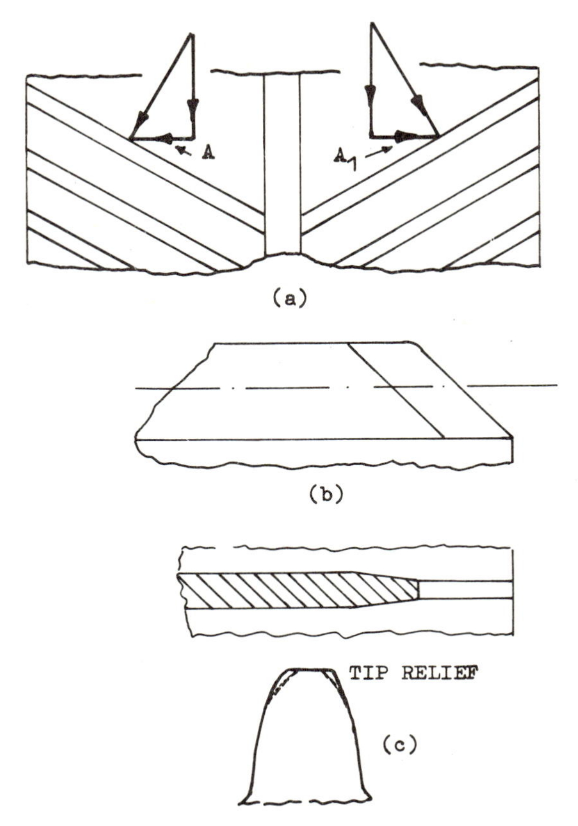

Fig. 44 Double helical gear

taught string as it is unwound from a base circle as shown in Fig. 45(a). The involute form gives a strong root section, improving the resistance to bending stresses whilst being able to tolerate a degree of misalignment without serious effect.

If two plain cylinders of different diameters are fixed to their respective shafts and placed in bearings so that they press hard—one against another—then when one is rotated, the other will turn in the opposite direction at a speed proportional to the respective diameters of the cylinders. At light loads there will be no slip between the two cylinders but as the load is increased, so one will tend to slip across the surface of the other. An important point is that under the no slip condition the linear velocity of the two points in contact will be the same. To prevent slip occurring and to allow high loads to be transmitted, teeth are formed around the circumference of the cylinders and many of the basic definitions involved in gear tooth design are related to the circumference of two such cylinders. The circumference of the plain cylinder is known as the pitch circle of the gears and the radial distance from the tip of the tooth to the pitch circle is known as the addendum whilst the radial distance from the pitch circle to the root circle is called the dedendum (Fig. 45(b)).

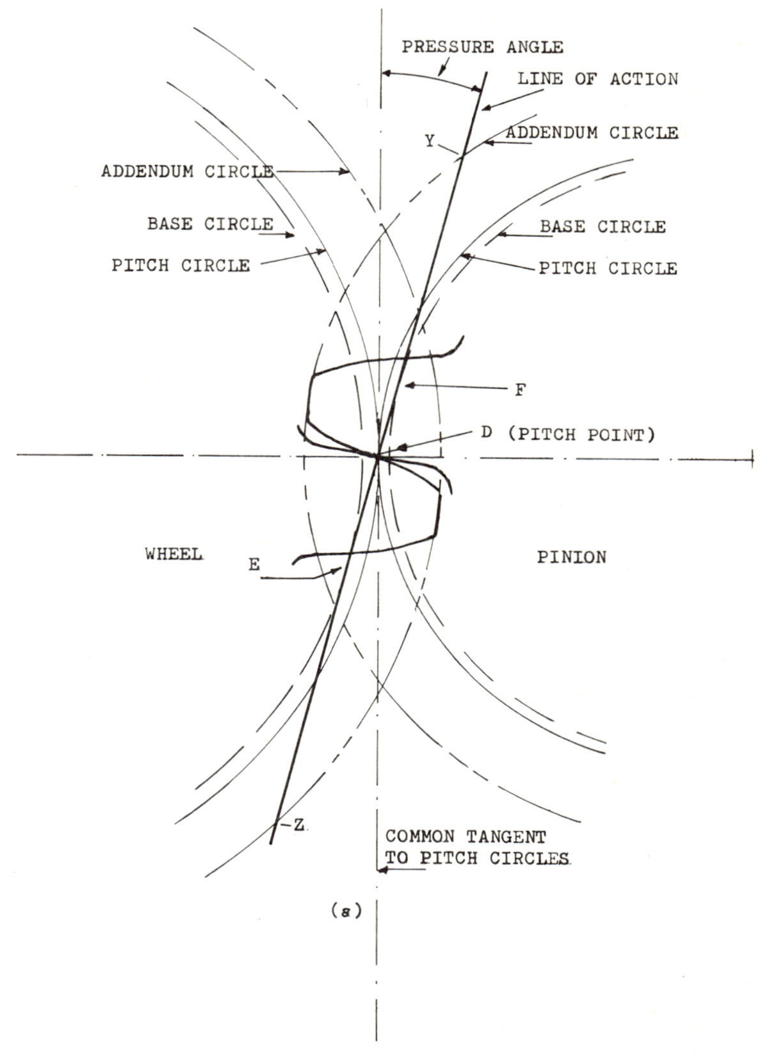

Fig. 45(a) Gear tooth design

The circular pitch is the distance measured around the pitch circle from a point on the flank of one tooth to a similar point on the flank of the next tooth.

When two teeth come into contact with one another at the pitch point D (the point where the two circles meet), ED is the radius of curvature of the flank of the wheel tooth formed from the wheel base circle, whilst FD is the radius of curvature of the pinion tooth flank formed from the base circle of the pinion. These two lines are tangential to both circles (the curve for the tooth flank is formed by unwrapping a

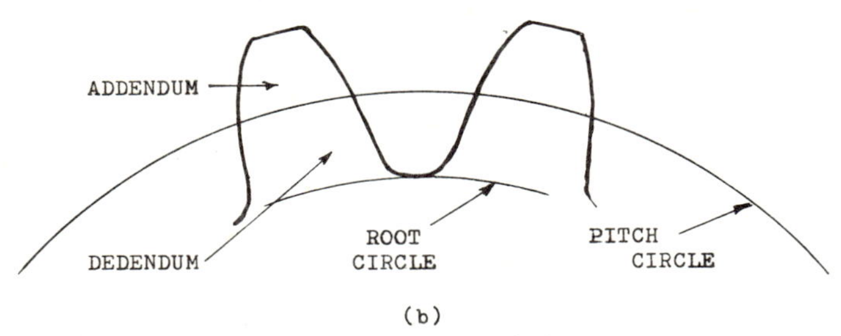

Fig. 45(b) Gear tooth definitions

taut string from a cylinder, ED and FD being the strings, the cylinders being the respective base circles). EF is thus the common tangent to both circles and passes through the pitch point. As it is the common normal to both surfaces it gives the direction of pressure between the teeth and is called the line of action.

If the path of the point of contact of the two mating teeth is traced from the moment of engagement to disengagement, it will be found for all positions during contact to be along the line of action EF. The length of the path of contact for mating teeth along this line is given by the intersection of the line and the addendum circles at Y and Z.

The pressure angle is the angle at any point along the path of contact between the common normal at the point of contact and the common tangent to the two pitch circles. It indicates the direction in which pressure is exerted between two meshing teeth. The angle is usually between $14\frac{1}{2}°$ and 20°, for if the angle is too large it tends to produce sharp pointed teeth requiring increased pitch, whilst if it is too fine it produces undercutting.

Gear systems

Fig. 46(a) shows the layout of a single tandem gear. The arrangement tends to produce a long gear case and the weight is somewhat high, but it has numerous advantages which make it a very common gear arrangement for marine work. The design can be used to transmit very high powers, whilst the length of the shafting provides a degree of damping to vibration, reducing its susceptibility to varying propeller torques and to minor errors in manufacture, alignment and to wear. Also primary and secondary reductions can be dismantled independently and there is freedom of choice with regard to pinion and wheel diameters. There is always a flexible coupling between the rotor and the primary pinion to prevent axial expansion or contraction of the former placing an end thrust on the double helical gears and to allow these freedom of axial movement, whilst there may be a similar coupling between the primary wheel and the secondary pinion. This sometimes leads to the term 'articulated' being applied to the design and the arrangement is used to provide increased flexibility, absorbing vibration and misalignment and preventing excessive wear due to overloading.

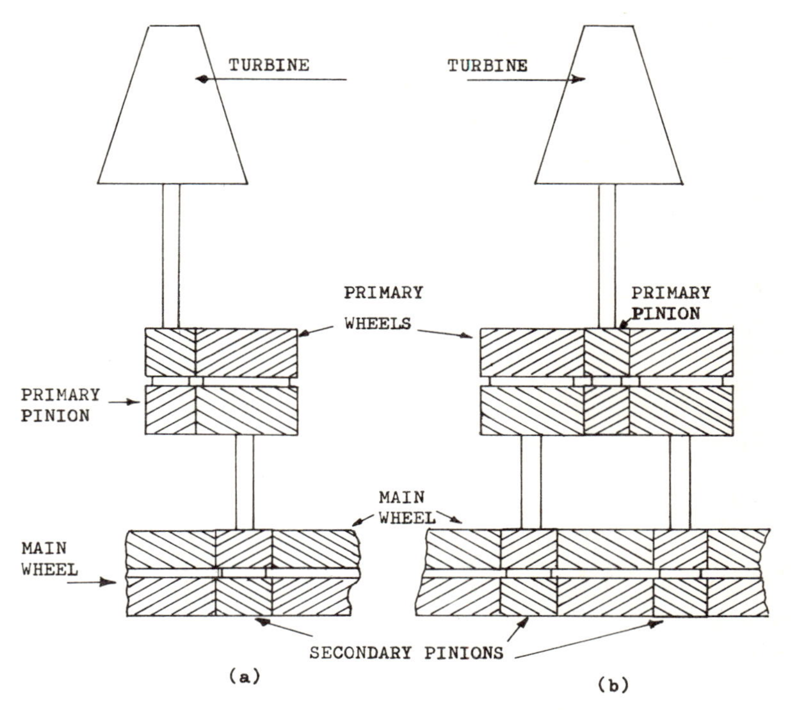

Fig. 46 (a) Single tandem gear; (b) Dual tandem gear

With the locked train design the power from each turbine to its primary pinion is equally divided over two intermediate shafts for transmission to the main wheel. This design enables a considerably smaller main wheel to be used and the overall gear arrangement is smaller and lighter than for the single tandem design. Again there is always a flexible coupling between the rotor and the primary pinion and usually between the primary wheels and secondary pinions. The design also incorporates quill shafts between the rotor and primary pinion and between the primary wheels and secondary pinions. Quill shafts and flexible couplings have to be used as the slightest pitch errors in gear trains solidly coupled to one another would lead to unacceptably high dynamic loads on the teeth. The quill shafts act as torsion springs resisting and making harmless these dynamic effects whilst maintaining uniform distribution of the power. By having two secondary pinions on the main wheel, greater use is made of the main wheel tooth area available for the transmission of power, allowing higher outputs for a given size of train without the need for high tooth loadings or expensive materials and manufacturing procedures. The exception is the primary pinion which is indeed very heavily stressed and great care has to be taken with the alignment of this and the primary wheels to ensure an even distribution of power. Fig. 46(b) shows the gear arrangement. Disadvantages include the large number of rotating parts and bearings, whilst there is only a small difference in height between the turbine pinions and the main shaft, producing a slanting propeller-shaft if an underslung condenser is used.

Typical gear ratios for a double reduction geared turbine are as follows:

HP primary 6·58 : 1
LP primary 4·11 : 1
HP and LP secondary 8·68 : 1

Number of teeth:

HP primary pinion 40
LP primary pinion 64
HP and LP primary wheels 263
HP and LP secondary pinions 65
Main gear wheel 564

None of the gear ratios is a whole number to ensure that all the teeth in one unit mesh in turn with all the teeth on the mating unit; in this way cyclic irregularities are prevented.

Gear defects

Pitting is usually found in its most severe form about the pitch line but may also be present in the dedendum of both the driving and driven gears and is dependent to a very large extent on the finish of the gears. New gears will invariably suffer some pitting, known as incipient pitting, but with careful operation this should disappear as the surfaces of the teeth work harden, and normal wear will polish out the pits. If, however, the tooth surfaces are poor or overloading occurs, pitting will proceed, gradually reducing the load bearing surface so that eventually the gear teeth are destroyed. This is known as spalling and may result in steps being formed in the gear teeth. It is caused by local high spots carrying excessive loads as the mating teeth come into contact, causing localized high temperatures which eventually cause fine fatigue cracks to occur. Subsequent loading tends to close these cracks and the entrapped oil is raised to a very high pressure which then breaks away part of the surface of the tooth. Careful honing with a smooth carborundum stone can help to relieve the trouble but care should be taken not to alter the form of the tooth flank.

Scuffing is caused by a local breakdown of the oil film between mating gear teeth as the surfaces slide across one another during engagement and disengagement. The disruption of the oil film causes local metal-to-metal contact, producing very high temperatures, resulting in fusion or welding of any high spots present. As the teeth slide across one another these small welds are torn apart damaging the surfaces. It is more prevalent at the tips and roots of the teeth as it is at these points where the teeth have a greater relative sliding velocity, and may progressively destroy the whole tooth surface, with bad scuffing tending to produce a feather over the tip of the tooth. Poor surface finish and overloading are the prime causes and it is most commonly found on the softer tooth surface (usually the wheel). A light honing and attention to oil viscosity and temperature may help to relieve the trouble.

Abrasion This is a form of surface destruction caused by abrasive materials entering between meshing teeth. It may appear in various forms, either deep channels are cut from the root to the tip of the tooth by a hard projection on one or more teeth piercing the oil film, known as scoring or ridging, or random scratching of the surface by small loose particles of grit or dirt being caught between meshing gear teeth. Another form produces a highly polished tooth surface due to very fine particles of dust being entrained in the oil. The only remedy is a high degree of filtration of the oil with a light honing to remove any high spots caused by the grooves.

Flaking This is caused by heavy overloading which overstresses the sub-surface of the metal. The heavy compressive stresses or shearing action on the sub-surface of the metal can exceed the yield point of the steel causing large flakes to break away. Rippling of the surface may also occur due to this defect, which may also be termed plastic flow.

Post hobbing processes

Provided gear teeth are properly designed, manufactured with care and subjected to the correct hardening and tempering procedures, if so required by the design, then if equal regard is paid to the loading, lubrication and operation, the gear teeth over a period of time should work harden to provide a smooth polished surface. As the surfaces become smoother, lubrication becomes progressively more efficient, friction and wear decrease until a condition is reached where further wear practically ceases. After a period of time there may be signs of long use, but the metal is rolled and polished to a smooth hard surface.

Shaving is a process carried out after a gear has been cut or hobbed and is commonly called a post hobbing process. A cutter in the form of a mating helical gear is run along the length of the tooth face to remove any undulations, high spots and rectifying any variations in the helical angle between the pinion and wheel left by the hobbing process.

Carburizing is another post hobbing process in which the pinion is held at a high temperature, 900°C, for a number of hours in a carbon rich atmosphere so that high carbon content gases diffuse into the surface of the teeth giving a hard wear resistant surface. Some growth of the tooth takes place and grinding may have to be resorted to in order to restore the required tooth form. Heat treatment is carried out after the hardening process.

Nitriding is a post hobbing process which involves exposing the surface of the pinion gear teeth to the action of ammonia gas at a temperature of about 500°C for a number of hours. Nitrides form on the tooth surface of the teeth providing a layer highly resistant to wear.

Materials for pinions and wheels

The properties necessary for a material for a gear tooth should be such that the tooth has a high fatigue strength to resist breakage and a wear resistant surface to withstand normal wear, scuffing and pitting. A common steel used for pinions contains 3½ per cent nickel and is hardened and tempered to a tensile strength of 618 to 772MN/m^2. The mating wheel would be a 0·30 per cent carbon steel having an ultimate tensile strength of 463 to 618MN/m^2, normalized and tempered. For higher powers a 2½ per cent nickel-chromium-molybdenum alloy steel would be used for the pinion, mating with an 0·40 per cent carbon steel wheel. (The 2½ per cent refers to the nickel content of the steel.)

It has been found that the resistance to pitting and the fatigue strength of a gear material increase with the tensile strength, but due to the shape of the tooth there is no gain in fatigue strength if the strength of the steel is taken to extremes. Nickel, chromium and molybdenum may be added and heat treatment carried out together with tempering, to raise the tensile strength of steels, and such methods are used to improve behaviour under increased loading conditions but the process would not give the required degree of pitting resistance for very heavily loaded gears such as some pinions have to withstand. Post hobbing processes such as carburizing and nitriding may therefore be used to improve the fatigue strength of the steel by giving a very high strength layer of metal over the tooth surface supported by a tough core.

The pinion is the weaker unit in a gear train as it suffers greater stress fluctuations and is more likely to fail from fatigue. The strength of a pinion wheel combination is therefore dominated by the pinion. To overcome this problem for speed reducing gears the pinion may be given positive addendum modification to increase the thickness of the root and thus improve its bending strength. It is used with pinions having small numbers of teeth to avoid undercutting during manufacture which weakens the root by cutting the metal away at this vital point and also reduces the working surface area over the length of the tooth. The all-addendum tooth form is a tooth design that has received the maximum amount of correction possible so that the tooth form is displaced outwards, bringing the root circle nearer to the base circle with only the root circle below the base circle. All the profile above the base circle is then of the involute form and the mating gear wheel teeth are then all-dedendum. The teeth engage with a pure rolling action at the pitch circle and are only in contact during the arc of recess, with the relative sliding in one definite direction over the whole of the tooth surface. It is claimed that it tends to give smoother running and reduces pitting, whilst the root thickness is greatly increased, reducing bending stresses.

Toothed flexible coupling

Each turbine is connected to its primary pinion through a fully flexible coupling which is designed to nullify the effects of slight misalignment between the rotor and pinion, to allow for the longitudinal expansion and contraction of the rotor and to accommodate the fluctuations in axial loading that can occur in a pinion, due to fluctuations in propeller loading and/or irregularities in the surfaces of the gear teeth. All these effects can produce forces which can result in excessive gear or rotor thrust wear if not absorbed by some form of flexible coupling between the two components. Fig. 47 shows a fine tooth coupling attached to the end of the turbine rotor shaft and driving the primary pinion. The male claw is fitted to a taper on the rotor shaft with suitable keys and engages with the female part of the coupling bolted to the primary pinion. The male claws are forged from carbon steel whilst the mating female claws are of nitrided steel, the different materials being used to minimize local welding under high contact pressures. Teeth on the male claws are in the form of 20° pressure angle involute stubs and are barrelled along their length over the tips and along the flanks to provide a high degree of flexibility. The female teeth are of similar form but straight cut. The reduction of friction between the teeth is of vital importance and oil is sprayed into an annular ring from whence it is fed by centrifugal force to the male and female teeth.

In the arrangement shown here, the rotor shaft passes through a hollow pinion before engaging the coupling. This is known as a quill shaft arrangement and is designed to increase the length of the shafting in the turbine rotor/gearing layout to accommodate misalignment and provide increased torsional vibration damping properties. A similar design may be used between the primary wheel and secondary pinion for the same reasons. By telescoping one shaft inside the other in this manner the length of the shafting is effectively increased without adding any appreciable length to the space taken up by the turbine/gearbox layout in the engine-room. By doing this the gears can be brought to mesh at a point in the shafting system where

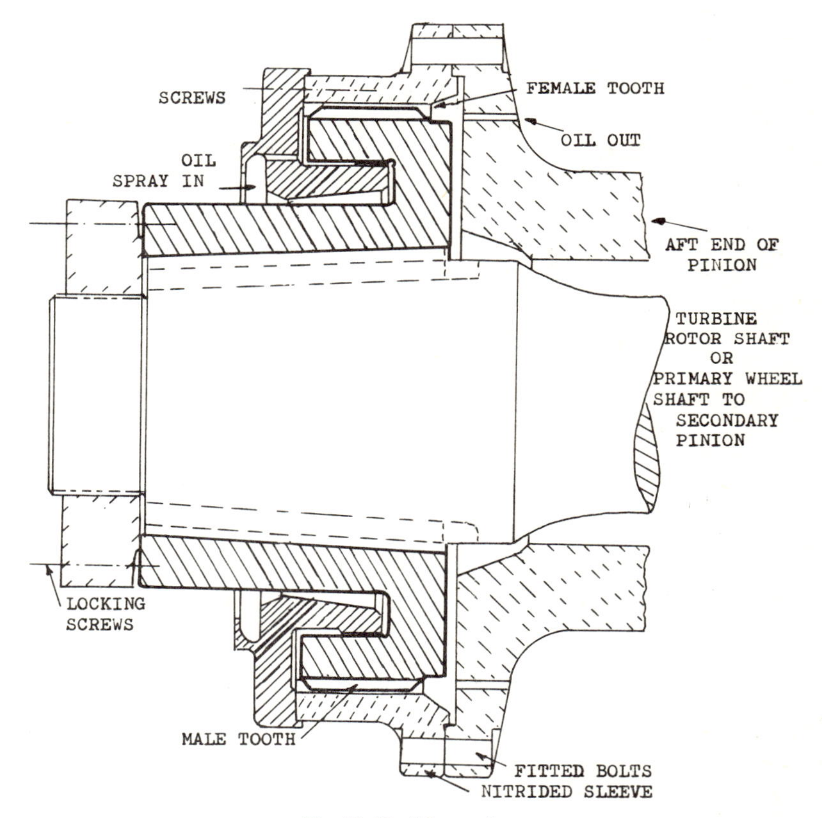

Fig. 47 Flexible coupling

there is little or no vibration so that the teeth mesh under a steady load instead of suffering cyclic fluctuations in load if they meshed at a point where vibration existed. This design is sometimes known as nodal drive: the point of no vibration being termed a node.

In some designs the flexible coupling is fitted between the turbine and the primary pinion, and in this case there are usually two sets of male and female claws, one set on the rotor and the other on the pinion. The two sets are then connected by a torque tube, for the transmission of the torque between the two couplings. They would be of similar design to the one shown.

It is very important that the greatest care is taken of the flexible couplings to ensure that they are free to move and not jammed with dirt from the oil supplied for lubrication and centrifuged out around the outer casing, or locked together due to fretting of the teeth setting up excessive friction.

Epicyclic gears: advantages etc

An epicyclic gear may be defined as a gear in which there is relative motion between the axes of the shafts such that at least one axis is capable of moving around another fixed axis, whereas in an ordinary gear train all the axes of the wheels are fixed. Fig. 48, (a) & (b) show the arrangement of the basic elements of

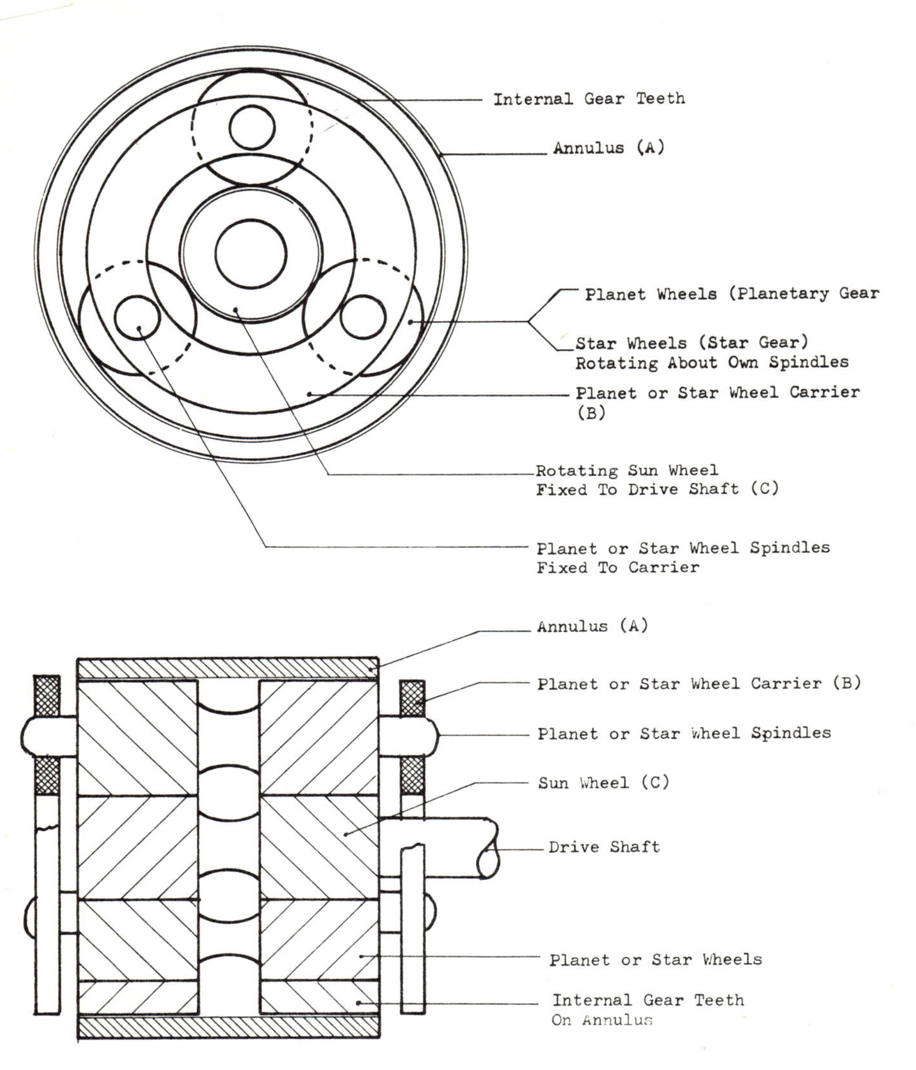

Fig. 48(a) and (b) Basic Components for Epicyclic Gears.

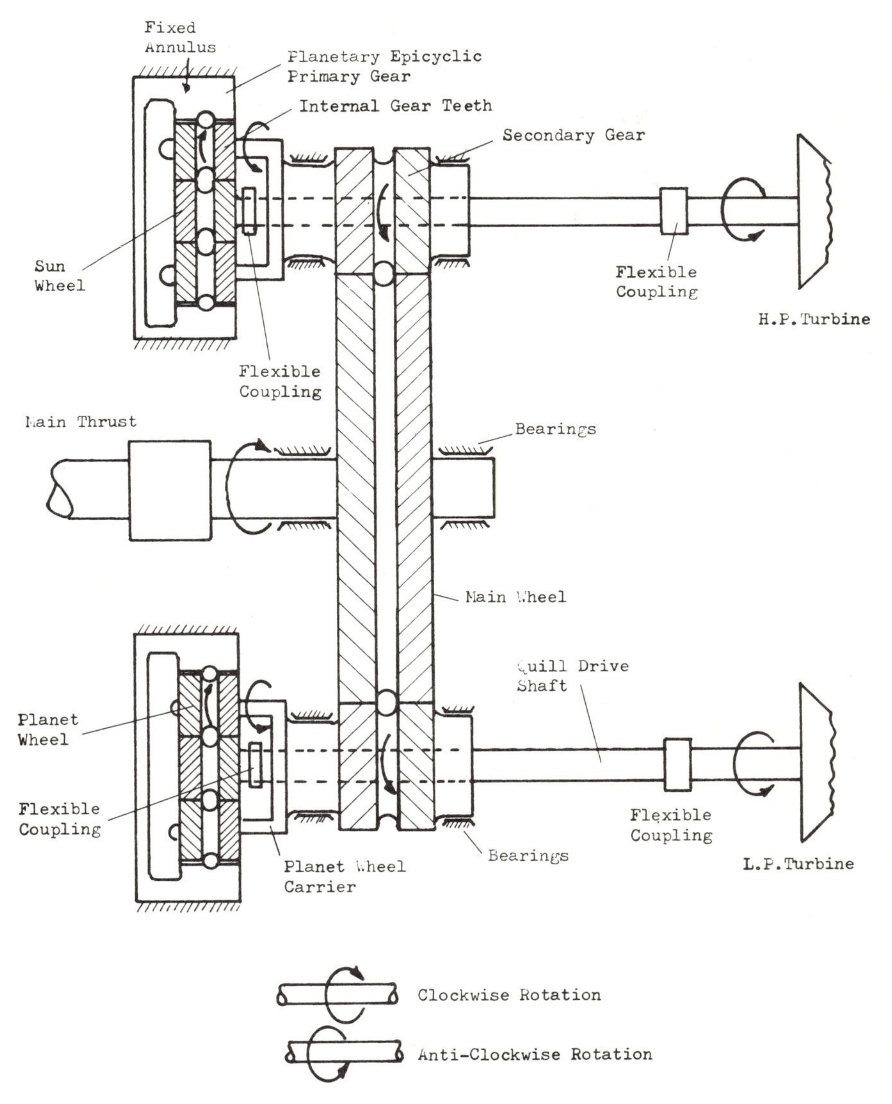

Fig. 49(a) Epicyclic Gears in Double Reduction Layout

such a gear, and these may be arranged so that any individual unit A, B or C, is held stationary. This has the effect of determining the greatest and the least ratios available and also the direction of the output shaft relative to the input shaft. In the planetary gear the annulus is fixed to the casing of the gear box and the planet carrier rotates in the same direction as the sun wheel. The gear ratio for this arrangement varies between 3:1 and 12:1.

With the star gear the planet wheel carrier is fixed to the gear box casing and the annulus rotates in the opposite direction to the sun wheel. This latter wheel is connected to the high speed shaft and the annulus to the low speed shaft. The gear ratio for this arrangement varies between 2:1 and 11:1.

It is claimed by manufacturers that the use of such gearing in one stage of a double reduction gear layout can provide a reduction in weight and floor area of approximately 30% each and a reduction of the machinery volume of about 25%. A very slight increase in efficiency is also claimed. It is also possible to obtain a reduction in machinery first costs and operating costs as well as allowing the turbine/gearing/condenser layout to occupy less height.

The sun and planet wheels are hobbed and ground and nitrided, whilst the parallel gear wheels and pinions are through hardened, hobbed, selectively shaved and positively nitrided. The planet wheels have their bores ground and polished with the spindles upon which they run being white metal faced.

Double reduction/epicyclic gear layout

Figure 49(a) shows the layout for a cross compound double reduction geared turbine installation incorporating single reduction epicyclic gears as the first reduction for both turbines, overhung from the final parallel reduction gear box. The epicyclic gear may be of planetary or star configuration, the former being shown. Where propeller speeds of the order of 80 r.p.m. are required it may be necessary to use triple reduction in order to accommodate a high speed H.P. turbine rotor and in this case a double reduction epicyclic gear is used for the primary reduction of this turbine, the first gear being of star configuration and the second planetary. The star gear is used when high power and speed output cause bearing loads to be too high on the planet spindles in a planetary gear. Note that the sun wheel and planet carrier are rotating in the same direction for the planetary gears.

On a conventional cross compound turbine layout a combination of epicyclic and parallel gears is required as the epicyclic gears are co-axial and the parallel gears are necessary to couple the turbines to a single propeller shaft.

The H.P. and L.P. turbine output power of Fig. 49(a) is transmitted to epicyclic primary gears via flexible couplings and fully floating sun wheels. The drive from each turbine is via a quill shaft through the pinion of the secondary parallel gears with the sun wheel rotating the planet wheels about their spindles. These are attached to a carrier ring which in turn drives the pinion of the parallel secondary gears. The annulus of the epicyclic gears is fixed so that the carrier ring rotates in the same direction as the sun wheel, and the parallel secondary pinion thus rotates in the same direction as the input. Each epicyclic gear box is overhung from the after end of the secondary reduction pinion to which the primary gear transmits power.

Figure 49(b) shows a triple reduction layout for an H.P. turbine. The output from this turbine is transmitted by a flexible coupling to the fully floating sun wheel of the primary epicyclic. This rotates the star wheels about their spindles, these being fixed by means of their carrier ring to the gear casing. The star wheels rotate the annulus which moves in the opposite direction to the sun wheel and in turn drives a quill shaft through the centre of the secondary reduction pinion to the second train of the epicylic gear. The quill shaft is attached to the fully floating sun wheel of this gear which drives the planet wheels about their spindles. These are attached to a carrier which rotates inside the annulus (fixed to the casing in this instance) and drives the secondary pinion of the parallel (secondary) gears in the same direction as the sun wheel. This secondary parallel gear pinion then drives the main wheel attached to the propeller shaft. The second train of the epicyclic gear is rigidly attached to, and its planet carrier overhangs from, the after end of the secondary reduction pinion.

The long quill shafts offer a degree of flexibility between the turbines and gearing for the absorption of slight mis-alignment and vibration.

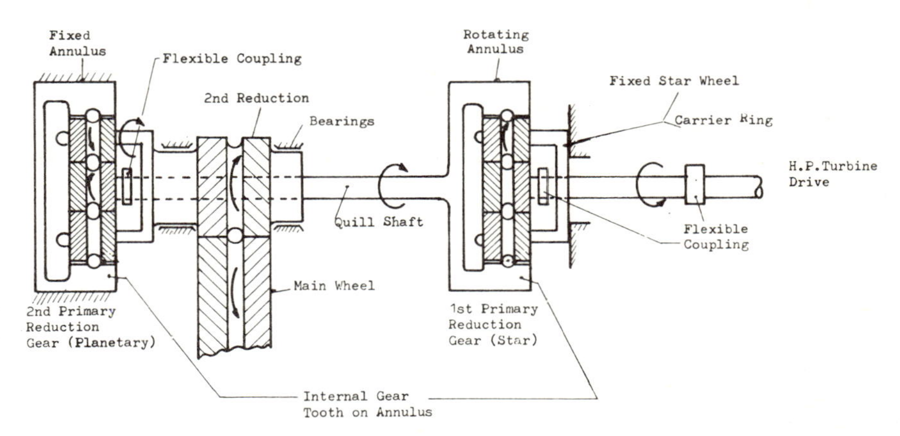

(b) Triple Reduction Gears

Note that the 1st primary reduction star gearing is used with the output shaft rotating in the opposite direction to the input shaft (sun wheel). For the primary secondary reduction the output shaft rotates in the same direction as the sun wheel.

Vibration monitoring

Continuously monitoring vibrations occurring in rotating machinery provides a means for detecting potentially dangerous defects at an early stage with the opportunity for a remedy to be effected before the damage becomes serious. Vibration measurements taken during sea trials establish the normal vibration behaviour of the plant for reference values to compare with future measurements, and in general significant changes in vibrations are more important to watch than absolute values. Changes in amplitude related to starting up and changes in load should be watched but not worried about to too great an extent once the normal pattern of the behaviour of the plant is known.

During normal operating conditions vibrations may increase slowly over a long period of time due to deposits forming on the rotating components or erosion causing irregular wear. Comparatively large vibrations may be permitted over a short period of time, but they should not exceed the manufacturer's recommended limits.

The most important feature to be watched for would be rapid changes in the vibration amplitude and deviations from normal behaviour of the plant. Such changes indicate serious deterioration in the operational conditions of the unit, even though the amplitude may still be within the permissible range. Defects causing this behaviour could be heavy blade deposits, blade failure, bent rotor or damaged gear teeth. Coupling misalignment, particularly after overhaul, can also cause an increase in vibration behaviour.

Points to watch for are: the length of time the high level of vibration has been in existence, whether only certain components are affected, does it vary with operating conditions or is it exceptionally high at one particular speed, does it occur directly after starting up, are slugs of water carrying over from the boiler the cause, or could casing drains be restricted, might high oil temperatures be the cause. Also, how long has the machinery been in operation before vibration recording was commenced.

Diaphragm couplings

In place of the toothed coupling between turbine rotors and gears which had problems with friction, tooth wear and lubrication, diaphragm type flexible couplings as shown in Fig. 50 are now being used. These allow for angular, lateral and end float by the deflection of laminated diaphragm plates. Due to the absence of sliding parts these couplings require no lubrication, and possess a high amount of axial and angular freedom. The stress set up in the individual diaphragms is low enough not to cause fatigue problems.

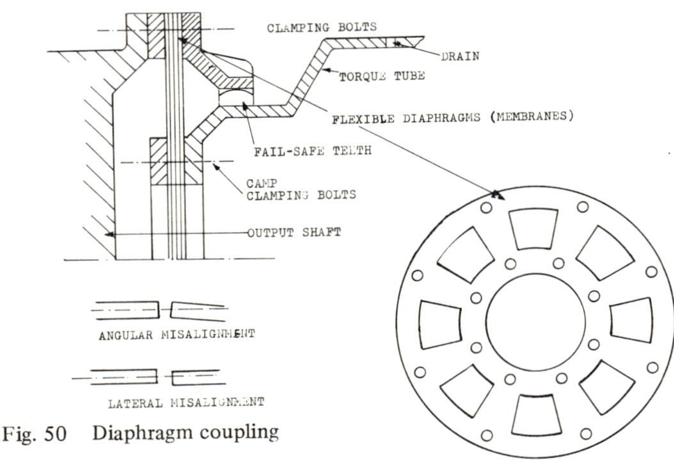

Fig. 50 Diaphragm coupling

It may be necessary to maintain drive under heavy overload conditions, and this can be done by incorporating splines (or teeth) which normally remain separated, only coming into use at overload conditions or in the event of sudden shock causing failure of the diaphragm plates.

Gear dehumidifier

In order to prevent gear corrosion caused by water contamination of the lubricating oil, a dehumidifier is frequently installed on the gearbox. It is intended to supply the gear with dry air under all operating conditions and gives the best performance when run continuously. The rotor is divided into two sections in the rotor casing with fans drawing in air from the atmosphere and from the gearbox. The atmospheric air is heated electrically and then passed across the rotor vanes. These are built up of flat and corrugated strips of asbestos paper, creating a large number of passages through which air can flow. A very large contact area is presented to the air flow in this way. The rotor is impregnated with non-inflammable hygroscopic material which is the actual absorption medium.

The rotor rotates slowly inside the casing at about 7 rpm during which time the portion of the rotor which is exposed to the air from the gearbox absorbs moisture from this air, which is then passed back into the gearbox. The section of the rotor exposed to the air from the atmosphere which has been heated, releases the moisture absorbed from the gearbox air to this atmospheric air which is then passed back out. The general arrangement is shown in Fig. 51.

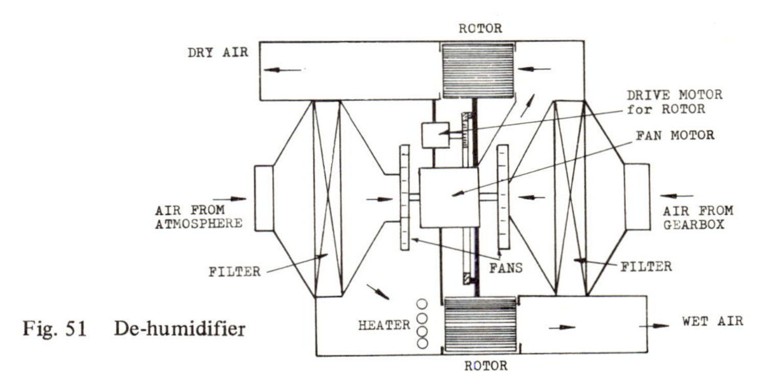

Fig. 51 De-humidifier

Gear inspection

It is generally recognised that the periods between the inspection of a set of gears should be six months or more unless there is suspicion of a serious defect. This is due to the danger of contamination that can arise at each inspection, and the more frequently the gearbox is opened up the greater the risk.

Before embarking on an inspection the records should be checked for information about any previous damage or wear noted so that comparisons can be made. The persons carrying out the inspection should make sure there are no loose articles in their pockets, particularly overall breast pockets, in case something should fall into the gearbox accidentally. When opening up the gearbox there should be no naked lights in the vicinity as explosive conditions can be generated in gearboxes.

For recording defects the easiest and cheapest way is to clean the tooth involved thoroughly and smear with marking blue. This is then cleaned off and a strip of Sellotape or Scotch tape placed over the tooth surface and firmly pressed onto it. The tape is then pulled off and placed on a sheet of stiff white paper, whence the defects will show up clearly.

For deep or extensive defects a cast can be made using plaster of Paris, which tends to shrink on hardening, or silicone rubber and resin moulds can be used. Silicone rubber produces particularly good results as it gives highly accurate details of defects such as machine marks, surface flow and micro-pits with exceptionally fine detail. It is possible then to produce a positive cast and aluminize this to give a very accurate detail of the gear surface.

Another procedure is to use acetobutyrate plastic film, known as Triafol, which is applied after flooding with dimethyl ketone and allowing the film to

..y. Slides can be prepared from this for projection or it can be aluminized for microscopic examination.

Photographs are useful for recording defects but shadows cast by the lighting can lead to discrepancies when making comparisons.

Gearbox gravity tanks

Fig. 52 shows the gravity tank arrangement used in Stal Laval lubricating systems. Oil from the system (1) flows through the orifice (A) in the plate in the

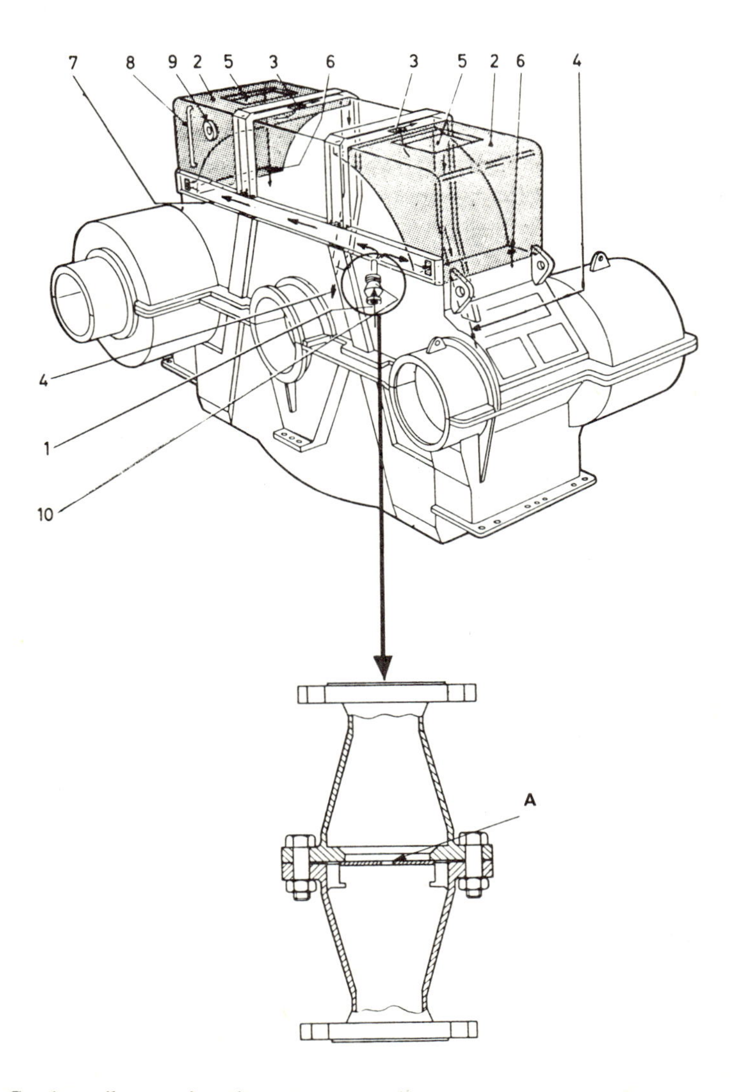

Fig. 52 Gearbox oil reservoir and emergency supplies

valve (10) into the gravity tanks (2) in the wings of the gearbox. In the event of pump failure the orifice plate drops down on to the lugs projecting from the plate guide. Oil can then flow down through a series of slots around this guide to feed the bearings and gears. The gravity tanks are above the level of all lubrication points at a level sufficiently high above the bearings to ensure an adequate pressure head in the event of pump failure. Excess oil during normal operation returns to the system through overflow ducts (3) and passage (4) via the bottom of the gear casing from whence it is returned to the separate oil tank. (5) is an inspection manhole and (6) is a hole for continuous drainage. (7) and (8) are sampling point and level indicator respectively. (9) is the connection for a level switch and (10) is the combined non-return and throttling valve with the orifice plate.

Fig. 53 shows the oil system run-out condition. Lloyds Register quote a period of six minutes for their requirements.

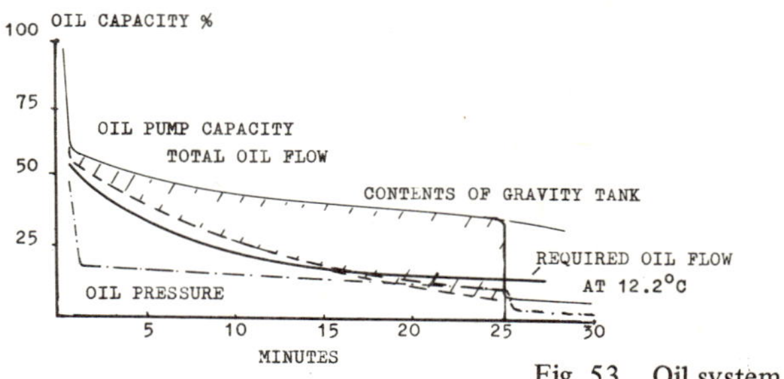

Fig. 53 Oil system run-out

Direct-driven oil pump

The direct-driven oil pump (Fig. 54) works in parallel with the electric pump. It is reversible in order to be effective even during astern operation. The pump consists of three screws, the threads of which are so formed that they seal against each other as well as against the casing. When the screws rotate, the seal between the threads moves axially in a perfectly uniform manner and thus acts as a piston moving constantly in the same direction.

1. Gear wheel, driven by a pinion connected to the second reduction of the reduction gear.
2. Spline coupling which transmits the rotation of a gear wheel (1) to driving screw (3).
3. Driving screw, guided radially by the casing (5) and axially by two thrust bearing rings (a) and a thrust bearing washer (b).
4. Driven screws, guided radially by the pump casing (5) and axially by the thrust washers (1) against the pump casing.
5. Pump casing containing three bores in which the driving and driven screws rotate. A silencing groove (c) provided in the driving screw bore releases the air entrained in the oil. This eliminates hammer in the oil pipes.

6, 7. Valves which are open during ahead operation of the machinery.

8, 9. Valves which are open during astern operation of the machinery.

10. Relief valve which discharges the oil back to the suction side of the pump when the discharge pressure is excessive. The valve is set and sealed during component testing at the works.
11. Air vent valve (only 1200 L/min. pump).

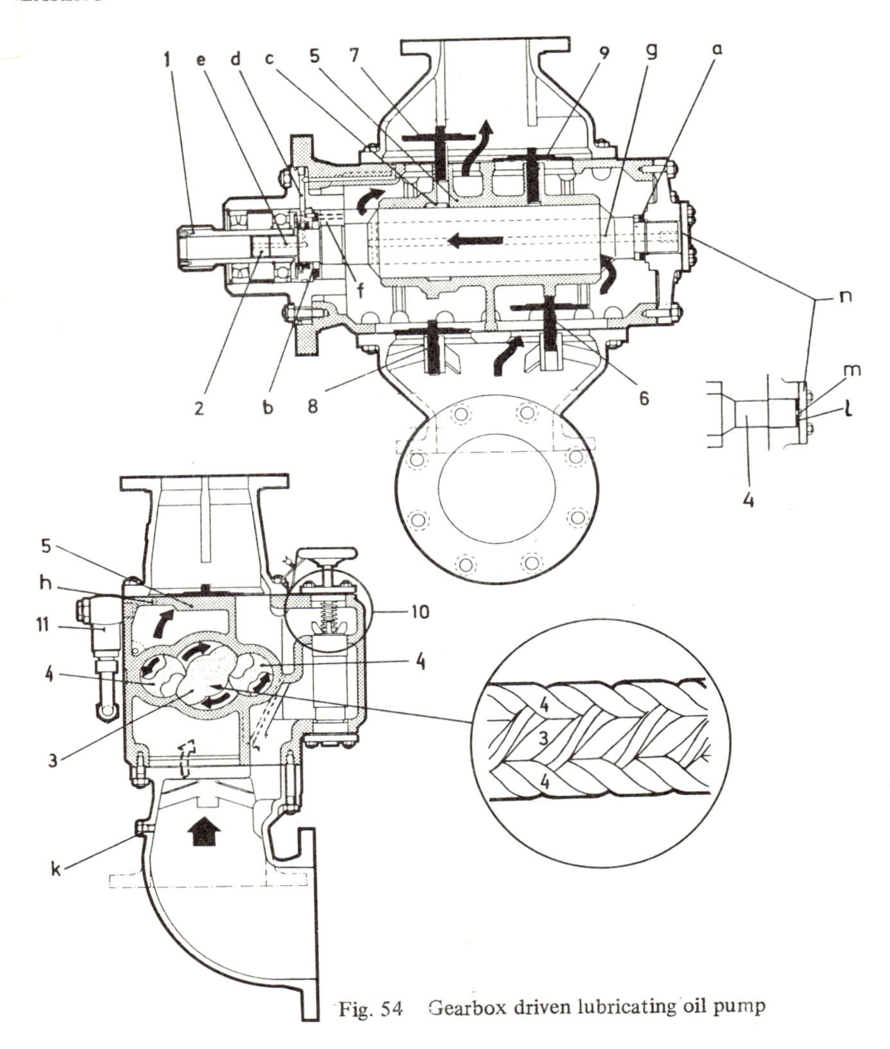

Fig. 54 Gearbox driven lubricating oil pump

Internal lubrication of the pump

The spline coupling is lubricated from the discharge side of the pump through hole (d) in the casing and (e) in the driving screw. The thrust bearing washer (b) is lubricated through hole (f). The thrust washers (l) are supplied through a hole (g) in the driving screw and hole (m) in the cover (n). Holes (h) are provided in the casing (5) through which oil flows from the discharge side, to ensure adequate lubrication of the screws during starting. (If the pump has been inoperative for a long period, oil is filled through the holes (i). Only 600 L/min. pump.) On dismantling the pump, oil is partly drained through the hole (k).

Gear materials

C%	Si%	Mn%	Cr%	Mo%	V%	Ni%
Sun wheel, planet wheel, star wheel			GK3 Nitralloy			
0.26 - 0.34	0.15 - 0.35	0.4 - 0.7	2.3 - 2.7	0.15 - 0.25	0.1 - 0.2	
			Tensile strength 1100-1200 N/mm^2			
			Hardness 335-365 HB			
Parallel gear pinion						
0.3 - 0.5	0.2 - 0.4	0.35 - 0.5	2.5 - 3.3	0.4 - 0.7		0.4 max.
			Tensile strength 780-890 N/mm^2			
			Hardness 250-280 HB			
Parallel gear wheel rim						
0.22 - 0.29	0.1 - 0.4	0.5 - 0.8	0.9 - 1.2	0.15 - 0.25		0.3 max.
Phosphorus and sulphur content of the above steels is 0.025% each max.						
Epicyclic gear inner gear ring			Tensile strength 860-1000 N/mm^2			
0.36 - 0.44	0.1 - 0.35	0.45 - 0.7	1.0 - 1.4	0.2 - 0.35		1.3 - 1.7
			Hardness 248-302 HB			
Planet carrier, for epicyclic gears						
Nickel	Molybdenum	Cast iron	Hardness 180-250 HB			
3.0	2.0	0.7		0.5		0.08

Index